BEEKEEPING

A step-by-step guide to setting up and maintaining a hive

ALICE MACKENZIE

Kerry County Library
Withdrawn from Stock
Withdrawn from Stock

This edition published in 2011 by Arcturus Publishing Limited
26/27 Bickels Yard, 151–153 Bermondsey Street,
London SE1 3HA

Copyright © 2011 Arcturus Publishing Limited

All rights reserved. No part of this publication may be reproduced, stored in a retrieval system, or transmitted, in any form or by any means, electronic, mechanical, photocopying, recording or otherwise, without written permission in accordance with the provisions of the Copyright Act 1956 (as amended). Any person or persons who do any unauthorised act in relation to this publication may be liable to criminal prosecution and civil claims for damages.

ISBN: 978-1-84837-751-6
AD001678EN

Printed in Singapore

The Fell Types are digitally reproduced by Igino Marini. www.iginomarini.com

CONTENTS

INTRODUCTION

There are several good reasons why people are becoming increasingly interested in taking up beekeeping. First and foremost there is the satisfaction of harvesting your own delicious, home-grown honey. However, more importantly, by caring for your own colony you will also be benefiting the ecology of our planet. Today, as a result of intensive farming methods, extreme weather conditions and a rise in bee disease, there is a rapid decline in the bee population across the world. As a result, our plants, including many of our food crops, are threatened – because, of course, as well as bees depending on plants, many plants depend on bees pollinating them in order to reproduce.

ANCIENT EGYPT

The first recorded mention of beekeeping dates from Egypt in 2400BC. The Ancient Egyptians believed that when the tears of the god Ra fell to the ground they were transformed into bees and flew away to pollinate flowers of every kind, thus creating wax and honey. Bees were kept in temple grounds in order to satisfy the gods' desire for honey and for the production of medicines and ointments.

The famous Smith Papyrus, an Ancient Egyptian textbook on trauma surgery, shows that honey was routinely applied to open wounds, making use of honey's antibacterial and fungicidal qualities.

Then there is also the sheer pleasure of relaxing in the garden in the sunshine and hearing the steady humming of bees at work among your flowers, one of the sounds most evocative of long and drowsy summer days. This book will show you how to become a beekeeper on a small scale, in your own garden. As you will find out, keeping bees does not require a lot of time, space or expensive equipment. Little intervention is required, as bees really only need our help when things go wrong.

However, there are certain basic tasks that must be attended to regularly, month by month. Early summer will be your busiest time and you will need to put by a few hours a week to make sure the bees have enough space to expand their brood and build up their supplies of honey.

With clear instructions and illustrations, this book will guide you through each step, offering sound advice to ensure your own personal safety, a healthy colony and a bumper harvest.

The structure of the book is broken down into seasons to help you plan ahead and gives you advice on what to do in any specific month. The calendar section starts in June as this is when, as a new beekeeper, you will have taken delivery of your first colony of bees.

The book is intended to encourage you to take the first step to keeping bees. Should you decide you want to know more, it is a good idea to enrol on a beginner's course that will show you all you need to know about this fascinating pastime. A variety of courses can be found through your local beekeeping association, or there are many now listed on the internet which may suit your purpose. Good luck with your new hobby.

THE GREATEST POLLINATORS

Without bees, the human race would struggle to survive. Many fruits and vegetables would become so rare that few people would be able to afford them. Quite simply, there would not be enough food to go around because bees do more pollinating than any other kind of insect.

Although there are thousands of different species of pollinators among flying insects, birds and bats, bees are by far the most prolific. The bee is perfectly designed for carrying pollen and even has baskets (called 'corbiculae') on its back legs specifically for this purpose. Each flower has its own special kind of markings which function like the lights on a runway, drawing the bees into the exact area where its pollen grains are stored. The bees' role in pollination is to move pollen from the anther (male part) of one flower to the stigma (female part) of another.

So why are bees so valuable to pollination?

- They are dispersed all over the world.
- Their bodies are designed with a fuzzy coating meaning that pollen sticks to them like glue.
- They carry a static electrical charge which also attracts the pollen.
- Bees collect pollen naturally as a way of survival.
- Certain types of bee tend to forage from a specific type of flower, which means it will pollinate more quickly.
- Because of their body design, bees can squeeze into certain tubular-shaped flowers (for example snapdragons), spreading pollen as they go.

PART 1

GETTING STARTED

TYPES OF BEES

It is impossible to list all the different species of bees as there are over 20,000 different types throughout the world. However, regardless of species, all bees are vital to the success of our planet as they are responsible for the pollination of much of our flora.

The bees listed below are the ones you are most likely to see in your garden.

THE SOLITARY BEE

As their name suggests, solitary bees are not inclined towards communal living. They spend much of their life as a pupa in a single nest cell and do not work with any other members of a colony. They do not make honey or wax and have a flying life of only 6–8 weeks. There are many different species of the solitary bee and they are very efficient pollinators, so you should encourage them by planting flowers in your garden. Unlike the honey bee, all the females in this species are fertile and once mated will lay around eight female eggs and one male egg in an underground burrow. Although they do possess a stinger, solitary bees are virtually harmless. They do not fly in swarms or spend their life guarding their nests, so they are not a threat to humans.

THE HONEY BEE

There are several species of honey bee, all belonging to the genus *Apis*, of which the most common 'domesticated' one is *A. mellifera* and its regional subspecies. Because they are so productive, these bees have been kept all over the world to pollinate crops and produce honey. Most popular with beginners is the Italian honey bee, *ligustica*, as it is relatively non-aggressive. Honey bees have a specially designed tongue (proboscis) that can suck up nectar (the sugary fluid secreted by plants) and their bodies are covered in a fine fur that traps the pollen (the fine yellow powder discharged by the male part of the flower). They live in colonies of up to 60,000 bees.

THE BUMBLEBEE

There are about 250 different species of bumblebee around the world. They can be easily distinguished from the honey bee because they are much rounder and have a thicker furry coat. Because of this coat they are able to fly in much colder weather and it is the bumblebee that you will probably see first in spring. They live in much smaller colonies – usually around 400–500 bees – and the queen will survive the winter by hibernating in a sheltered spot. They often nest underground in old rodent burrows or you may find them in cracks in walls or in leaf litter. They do produce honey, but only enough to feed their young. Unfortunately intensive farming has removed many of their natural habitats and this beneficial pollinating insect is now in decline.

THE WASP

Although wasps were the distant ancestors of bees, their evolutionary path diverged and they are now very different. Wasps are generally feared as they hang around picnic areas, dustbins or anywhere the smell of food attracts them, and unlike the honey bee they can sting more than once. They do not produce wax or honey, but some species are beneficial to your garden as they eat aphids, mealy bugs and other pests on your plants.

INSIDE THE COLONY

If you were to peer inside a hive of honey bees, you would see a flurry of activity and perhaps assume the insects were merely buzzing around between trips to collect nectar and pollen. In fact, inside the beehive is a hierarchy of roles, a highly organized system with each bee working towards the development and survival of the colony. This is known as the caste system.

LIFECYCLE OF A BEE

Inside a beehive is a wax honeycomb, built by the worker bees, consisting of a series of hexagonal shapes known as 'cells'. The queen bee searches for clean, empty cells in which to lay her eggs, backing into them and depositing a single egg in each. Using propolis (a type of bee glue – see pages 124–125), she attaches the egg to the rear of the cell and then moves to the next available cell to repeat the process. She can lay as many as 2,000 eggs in a single day.

The egg is similar in appearance to a grain of rice. After three days a larva will hatch and immediately require feeding. This is where the nurses (worker bees) play their part by providing the hatchlings with a substance called royal jelly, produced in their saliva. This is fed to the new larva for the first three days. Any larvae that are not destined to become new queens (and this is the majority of them) are then fed on a mixture of pollen and honey.

The larva goes through a series of different stages, which include shedding its outer skin

THE INTERNAL ANATOMY OF THE BEE

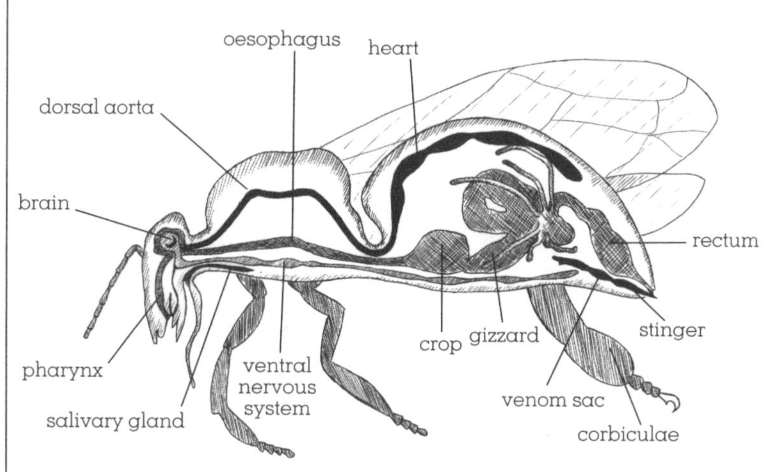

Bees are typical insects in that they have three body regions – the head, the thorax and the abdomen. Within these regions the following parts are located:

brain – the main engine of the bee's faculties.
corbiculae – the pollen basket.
crop – a storage area between the oesophagus and the gizzard.
dorsal aorta – a blood vessel in the back that takes blood from the heart to the other organs.
gizzard – (or *proventriculus*) a muscular extension of the crop.
heart – the organ that pumps blood round the body.
oesophagus – part of the digestive system, located just below the mouth.
pharynx – the intersection of the respiratory and digestive tracts.
rectum – the final part of the intestine.
salivary gland – where the bee's saliva is produced.
stinger – the bee's defensive mechanism.
venom sac – the pocket that contains the venom.
ventral nervous system – a collection of nerves in the abdomen.

several times. Each day the larva increases in size until it has to curl up to accommodate itself inside the cell. On the sixth day the larva straightens itself out and no longer requires feeding. The worker bees then cap the cell with a layer of wax and propolis and the larva is left to pupate. It does this by spinning a cocoon where it gradually turns from the milky white grub into a bee. The new bee will then eat its way through the cap and emerge as a brand new member of the colony.

For the worker bee this process will take 21 days, while the drone (see opposite page) emerges after 24 days. The queen develops much quicker and only takes 16 days to complete her cycle, having been fed purely on royal jelly which accelerates her development.

The early stages of a bee's life all take place in the part of the hive called the brood box (see page 35) and this is where the queen will spend most of her life. As a new beekeeper you will need to acquaint yourself with these various stages – egg, larva, pupa and adult – so that you can check the overall health of your colony.

The next stage of your initiation into beekeeping is to develop the ability to distinguish the worker bees from the drones and the queen herself.

THE WORKER

Over 90 per cent of the bees in the colony are worker bees (infertile females) and they perform almost every

The Worker Bee

duty within the hive. When the worker bee first emerges from her cell her stinger and wax glands have not yet matured and she starts her life as a cleaner – her job is to inspect the cells and make sure that they are in a fit state for the queen to lay her eggs in.

As she develops, the worker takes on the nursing role of feeding the new larvae with pollen and nectar. Once her mandibular and hypopharyngeal glands develop she can move on to feeding royal jelly to the queen, and to any future queens still in larva form. The next stage of her maturity is the development of her wax glands. This enables her to cap the brood cells and also to build and repair the honeycomb.

Once she is 14 days old, the worker is able to take pollen and nectar from the foragers, but she will not be capable of making her own trips until she is 21 days old. From 18 (when her stinger and mandibles have developed) to 21 days she acts as guard to the entrance of the hive. She develops a keen sense of smell which allows her to tell whether the returning foragers are from her own colony. If they are strangers she will attempt to drive them away. If they ignore her warnings, she will resort to stinging the intruder but will lose her own life in the process.

The worker bee's life is exhausting. Many only survive the first few weeks of their life, while others which are born later in the season can live through the winter as part of the cluster. This is the way in which the colony survives the colder weather (see page 79).

THE DRONE

In comparison to that of the worker, the drone's life is uncomplicated. The drones are the males and their prime role is to mate with a virgin queen. The drone does not have a stinger – this is replaced by the male reproductive organ. If they are not out searching for a queen, they spend their time in the hive waiting to be fed.

Drones are hatched from the queen's unfertilized eggs. Their development is not dissimilar to that of the worker bees, but at the larval stage a drone is fed for seven days before pupating and takes longer to develop than the female bees. Drones may occasionally assist the workers in ventilating the hive, which involves flapping their wings to circulate the air, but apart from that their role is minimal. Unlike the worker, the drone will sometimes move to a new hive in search of a virgin queen to mate with.

The Drone

The drone's life is short, as mating is the last action he will ever take. The drone and the queen mate in flight, and as the drone withdraws his sexual organ from the queen it is ripped from his abdomen and he falls to the ground and dies. Even if he does not mate and makes it through to the autumn, he will be ejected from the hive by the workers because they do not want to waste their stores of honey on him – meaning he will die when the weather turns cold.

THE QUEEN

The queen is aptly named, since out of approximately 60,000 bees in a hive, there is only one queen and she is the only bee to have developed

The Queen Bee

female reproductive organs. Her role is to lay eggs and spend the majority of her life in the darkness of the hive. She will only venture out once to mate or if the entire colony swarms (see pages 91–95).

Any egg that has been fertilized can become a queen bee if the colony decides they need to replace the existing one. The chosen egg will hang vertically in a queen cell (which is larger than a normal cell) as opposed to the other larvae, which lie horizontally. This larva will be fed solely on nutritious royal jelly approximately every five minutes for the first five and a half days. This will stimulate the larva to develop into a queen which, when she first emerges, is unable to fly and is sensitive to light. However, her first instinct is not to fly but to kill any rival queens, whether they are emerged or still pupating.

The mating flight

A couple of weeks after emerging, the new queen is usually ready to take her mating flight. Once out of the hive – usually on a warm and calm day – the queen will spiral up to find a swarm of drones waiting to mate with her. She will mate with about 15 drones, or until her sperm sac (spermatheca) is completely full. The queen will then return to the hive and spend the remainder of her life laying eggs.

Scent and sound

The queen has a particular odour which is passed to every member of the colony by mouth when they are grooming her. This scent tells the workers whether the colony is healthy. As the queen ages her odour changes, which indicates it is time that she is replaced. The workers will raise a new queen which, when she emerges from her cell, will kill the old queen and take over her role, a process which is called supersedure. Alternatively, the worker bees may kill the old queen by clustering round her so tightly that she dies from overheating.

In addition to her scent, the queen uses different sounds to communicate with the rest of the colony. When a new queen is ready to emerge from her cell, she will use a 'quacking' sound to let the workers know she is on her way out. If she wishes to challenge another virgin queen she produces a 'piping' sound. This sound also acts as a kind of defence mechanism if the old queen is attacked by workers. Both sounds are made by vibrating plates at the base of the wings.

To control the problem of competition within the colony, the queen produces 20 or more different pheromones that control the behaviour of the worker bees.

DEVELOPMENT OF A QUEEN BEE

Day	Stage
1 (egg laid)	egg
2	egg
3	egg
4	larva
5	larva
6	larva
7	larva
8 (cell sealed)	larva
9	pre-pupa
10	pre-pupa
11	pupa
12 (red eye)	pupa
13 (yellow thorax)	pupa
14 (yellow abdomen)	pupa
15 (pupa moult)	pupa
16 (queen emerges)	adult

HOW DO BEES COMMUNICATE?

THE ART OF DANCING

This form of communication is perhaps one of the most fascinating aspects of the behaviour of honey bees. This so-called 'dance' is only performed by a worker bee that has just returned to the hive after a spell of foraging. By using a particular set of movements, the bee can communicate to its fellow workers the exact distance and direction where the pollen is to be found, so that the others can go and collect pollen from the same place. The distance and direction are represented by separate movements within the dance.

THE 'ROUND' DANCE

If the food source is within 30 m (33 yd) from the hive the returning forager will perform a round dance by running around in small circles (see diagram below). During the dance the worker will change direction and may repeat the

The Round Dance

dance several times, either staying in one location or moving to other parts of the hive. Once it has completed the dance, the worker will often distribute a sample of the nectar or pollen to the bees that are in close proximity so that they can get used to the taste and smell. This will help them find the source of the food that the dancer is telling them about. The speed of the dance is also significant – the closer the food is to the colony, the faster the bee will dance.

THE 'WAGGLE' DANCE

To communicate the distance and direction from which they have come, bees vary their movements and start to perform the 'waggle' dance. This is used if the foraging site is more than 150 m (165 yd) from the hive. The bee will run straight ahead for a short distance, returning to the starting point in a semi-circle (see the diagram below). Then the bee turns a semi-circle in the opposite direction so that it has formed a

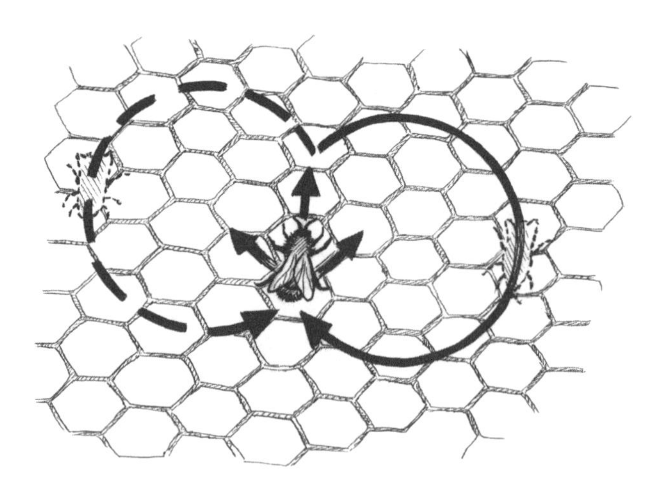

The Waggle Dance

complete figure-of-eight. While the bee is making these moves, it waggles its abdomen and tail from side to side, hence the name of the dance. At the same time the bee emits a low buzzing sound, which is produced by beating its wings at an extraordinarily fast rate. The straight move within the waggle dance indicates the distance the worker will have to travel to the food source. As the distance increases, so too does the duration of the waggling portion of the dance. For example, if the bee's waggle run lasts for 2.5 seconds, this indicates that the source of food is approximately 2,500 m (2,735 yd) from the hive.

The direction the other worker bees have to travel is indicated by the waggle. If, for example, the source is 40 degrees to the left of the sun, the waggle run will be 40 degrees left of the vertical. This will be repeated so that the other workers are able to familiarize themselves with the movements and find the source of the food.

THE USE OF SOUND

In addition to the movement the bee will also emit certain sounds to guide its fellow workers. This is vital as the inside of the hive is dark and consequently sound and vibration are key to the success of the communication. To try a sample of the nectar or pollen brought back to the hive, the bees can use a special 'stop' signal. They need to identify the smell of the pollen as this, interestingly, has a totally different scent to that of the flower from which it was taken.

The way in which bees communicate by dance is very important to their foraging and is so complex that it has only recently been understood with the help of modern technology.

COUNTY KERRY LIBRARY SERVICE

POINTS TO CONSIDER

Before making your final decision as to whether you would like to keep bees or not, there are several points you will need to consider.

- Do you have space in your garden or in an area such as a roof terrace where you can place a beehive?
- Is the site suitably sheltered from inclement weather?
- Do you have somewhere to store your equipment?
- Could a beehive pose a problem to your immediate neighbours?
- Do you have enough spare time to devote to your new hobby?
- Do you have young children and pets in your family?
- Are you willing to plan holidays around your hobby?
- Can you afford the costs involved in setting up?

SPACE

Although you might assume you need a lot of space to keep beehives, this is not necessarily the case. The hobby is now becoming more and more popular with people who live in cities, as a hive can be placed in a small courtyard garden, on a roof, or even on a balcony. Because the numbers of honey bees in more traditional locations have fallen dramatically in recent years, city dwellers can play a vital part in helping to reverse the downturn.

As long as city dwellers arrange their space to encourage bees, there is no reason why they should not be kept in urban areas. In fact a new, easy-to-maintain beehive, the Beehaus, has been introduced on the market to encourage city apiarists. The Beehaus is a safe, modern hive that makes beekeeping straightforward and comes complete with everything you will need, right down to gloves, smoker and the bees themselves. Because this plastic hive does not have all the nooks and crannies of the wooden ones, it is not so prone to disease. Even if you have only a balcony or some roof space, why not contact your local beekeeping association and find out more about keeping bees in the city? Many of these associations run courses which will give you more idea of what keeping bees in your particular situation involves.

The main concern to urban beekeepers is making sure that your bees do not become a problem to neighbouring properties. If you have little space it is probably best that you keep only one or two hives to lessen the risk of annoying the neighbours. Keeping bees in the city is a wonderful way of bringing nature to urban gardens.

EQUIPMENT

One of the main points to consider is whether you have somewhere suitable to store your equipment (see pages 37–41). This space can be an outside shed, a spare room, or perhaps a basement. If possible it should be free from damp and somewhere that pests cannot penetrate.

POSITIONING THE HIVE

Finding the right position for your hive is very important, as you want your bees in a convenient place and yet not too close to your house or that of your neighbours.

Bees love the sun first thing in the morning, so try to find a spot that catches the early rays. If the hive is warmed up by the sun early in the day this will encourage the bees to go foraging, which will mean a bumper harvest for you. Consider the following points when choosing where to site your hive:

- A main factor in siting your bees is how well they will be sheltered from prevailing winds, so find an area in your garden where brick walls, fences or thick hedges would offer your hive good protection. The hive also needs to have adequate shelter from the heat of the midday sun. Another bonus with thick hedges surrounding a hive is that they naturally encourage bees to fly upwards and over people's heads.

- Do not place your hive directly under a large tree as there is a risk of falling branches in bad weather conditions. If your hive is struck, it could badly damage it and cause either the death, or mass exodus, of your colony. Your neighbours won't thank you if there are suddenly thousands of homeless, angry bees on the loose!

- If your garden slopes, keep your beehive on the upper part. When the weather turns frosty the lower areas of your garden can be several degrees colder than a higher area. In this position rainwater will drain away downhill and your beehive will be less vulnerable to flooding or snow.

- Ideally, the site you choose for your beehive will have plenty of flowers nearby for nectar-foraging. However, do not place the hive too close to flowering plants as the bees often use the area directly outside their hive for their cleansing flights (see page 79).

- Position the hive close to where you store any additional equipment. You will not want a long journey carrying honey or supers (the frames which hold the honeycomb).

NEIGHBOURS

You may find that your neighbours will not be happy if they see you are putting a beehive in your garden because they will worry about the risk of being stung. Besides passing the occasional jar of honey over the fence, other ways to avoid complaints from next door are:

- Start with one beehive and see how you go. If your neighbours are not too bothered by the bees, you can always get another hive or two.

- If your neighbours are disturbed by the beehive, position it so it is not in plain view.

- Reduce the risk of robber bees attacking by reading the section on pages 90–91. If your hive is targeted and your bees survive, they may have become more aggressive in nature because of the attack, which is far from ideal if you live close to your neighbours.

- Ensure there is a water supply nearby for the bees – this way they will not have to visit a neighbour's pond too often.

- Explain the vital part that bees play in our food supply to your neighbours.

CHILDREN AND PETS

When your children are small it is important to teach them that bees are not dangerous – reading them stories about the famous Pooh bear and his love of honey will help with their general acceptance of this little insect. There will always be the risk that your child or pet may get stung, but if you take a few simple precautions you should not experience any major problems. If your children are very young, a fence around the outside of the beehive will keep them at a distance until they are old enough to understand about the colony and its activities.

Children are generally inquisitive by nature and teaching them how the bee colony works can be a fascinating subject. It is possible to buy bee suits in small sizes, so let your children become involved when you come to inspect your hives. Their eyesight is generally a lot sharper than that of an adult and you can use this to your advantage by asking them to spot the queen for you and to see if there are eggs in the brood box.

You can also get your children involved in extracting the honey and in making items such as candles or polishes from beeswax, which most children enjoy.

As far as pets are concerned, cats are usually sensible and

do not attempt to catch bees. Dogs, on the other hand, are far more interested and love to chase insects. It would be advisable to make sure they are shut indoors whenever you are carrying out an inspection of the hive.

As bees are passive creatures unless provoked, you should find that your family – including children and pets – can live quite happily alongside your bees.

TIME AND HOLIDAYS

You will need to put some time aside for your new hobby. The winter months require very little work, but you need to be prepared to put in a few hours once the honey is in full flow in the summer months. A weekly check is necessary at this time of year. A notebook with details of each trip to your hive will be helpful and will give you an idea the following year just how much time you will need.

You may need to think a little more carefully about when to take your annual holiday, but even if you decide to go away in the busiest season you can still plan ahead. As long as you have provided the bees with enough supers (see pages 35–36) it will be safe to leave them for 2–3 weeks as they will continue their duties without human intervention.

COST

Once you have purchased your initial equipment and colony, the cost of keeping bees is minimal. Borrowing equipment when it is time to harvest the honey can cut down on expenses – hives do not need to cost a fortune. See pages 31–46 on the different types of hives and equipment. You will also need to consider the cost of public liability insurance to cover against accidental injury and damage to property of third parties.

BEE STINGS

A bee will sting you if it feels threatened, has been roughly handled or has been stepped on with a bare foot. Drones do not have stings, so you cannot be harmed by them. Queen bees use their stings to kill other queen bees during the process of supersedure (see page 18), while worker bees will defend their colony by stinging anything they perceive to be a threat.

Although many people may have suffered a sting at some point in their lives, it is obviously those involved in beekeeping who are at greater risk. A bee knows instinctively that it must protect its colony and, in turn, the colony is protective of the individual bee. If you kill a bee it will emit a pheromone that tells nearby bees there is a predator close by, increasing the likelihood that you will be stung. This same pheromone is also released by bees under the following circumstances:

- If you smell strongly of perfume, hairspray or sweat.

BEE STINGING APPARATUS

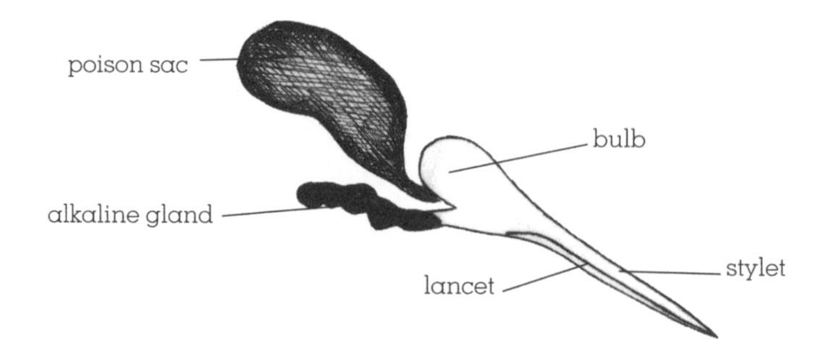

- If you make loud noises.
- If you are wearing dark clothes.
- If you are waving your arms around in a panic – bees become unnerved by the sudden movement and this may cause them to be on guard.

Larger numbers of bees will be alerted by these pheromones and it is in this situation that you are most likely to be stung. If bees seem agitated and fly at your face and around your head, and you experience a loud, high-pitched sound and vibrating sensation in your ears, chances are you have been targeted by the colony. This is certainly an unpleasant experience, but the pain is not severe and does not last a long time.

If you are stung, try to calmly brush off the offending bee. Most people would find it hard to stay calm in this situation, but after a few bee stings you will be less worried by it. If the bee is still on you, try to brush it off with a gentle slide of your fingernail or hive tool, and then remove the stinger. The easiest way to do this is to flick it out with the blunt edge of a knife rather than using a pair of tweezers, as this could leave the end of the stinger intact. Wash the wound once the stinger has been extracted. Most people experience a mild stinging sensation and a little swelling in the area, followed by a couple of days of mild itching.

In rare cases, some individuals may suffer a life-threatening allergic reaction (anaphylactic shock) to a bee sting, so some beekeepers carry a special adrenaline injector to use if a sting occurs. If you are at all concerned about a bee sting or you start to show adverse reactions you should always seek medical help urgently.

BEE STING REMEDIES

First make sure you have removed the stinger as quickly as possible, then try one of the following remedies to relieve the pain and reduce the swelling.

- Mix together vinegar, baking soda and lemon juice to form a paste and spread onto the affected area.
- Cut a piece of onion and place it directly on the site of the sting to minimize pain and bring down the swelling.
- Crush basil or parsley leaves and apply to the affected area.
- Applying toothpaste is one of the easiest methods and it is something you usually have to hand if you are stung at home. It contains alkaline substances and helps to neutralize the acidic effect of bee sting venom.
- Cool the sting down by putting an ice cube on it.
- Honey is one of the best cures for bee stings. Apply it directly to the area for temporary pain relief and to speed up the healing process. If you have a piece of gauze available, place it over the site of the sting to keep it clean. Honey works well because it not only has antibacterial properties, it also contains a certain amount of hydrogen peroxide which has an antiseptic effect.
- Make a paste out of bicarbonate of soda and water and apply to the affected area. Allow the paste to dry and leave on for about 20 minutes. Wipe off the paste with a damp cloth and if the area is still irritated, reapply it.
- Mix together two drops of tea tree oil and one drop of lavender oil to relieve the irritation.

A HOME FOR YOUR BEES

In the wild, bees will live in a dark crevice away from the damp and with an entrance which is small enough to keep predators out. Using wax that they have produced themselves, they will build a set of interconnecting cells in which to lay their brood and to store their honey. The distance between each sheet of comb is very important to ensure that the bees can pass one another without getting crushed, but is not so large a gap that they lose bodily contact (see page 72).

We humans can also provide a comfortable home, or hive, for bees and the next few pages explain each individual part of a hive and how the bees utilize it. There are several different types of beehives on the market; the most popular and widely used hives around the world are the Langstroth and the National.

Most bee suppliers will be able to provide you with a beginner's package that will offer you everything that you will need to start. Buying second-hand equipment is also an excellent way to start beekeeping. However, make sure it comes from a reputable source, that it is in good condition and that you treat it thoroughly for any possible disease – particularly American foul brood (see pages 133–134).

HOW A HIVE FUNCTIONS

Although it might seem rather complicated to the novice beekeeper, a hive is really just a simple set of individual boxes stacked together to make the complete unit. The following pages explain all the components and the purposes they serve in housing a successful colony. The true art of successful beekeeping is knowing how many frames and boxes to put on the hive, and the appropriate time to add more or remove them. You will need to learn exactly how to read what is happening on each frame and this will be discussed in detail further on in the book. Starting at the bottom of the hive, here is a description of each part of the hive and the role it plays:

THE STAND

Designed to keep the hive off the ground, the stand has to be a sturdy structure. It needs to be high enough to prevent damp and unwanted insects entering the hive from below, and also to make sure that the bees' entrance to the hive is not blocked by grass or weeds. An ideal height is 76 cm (29 in). The stand can be constructed from wooden blocks, bricks or even a small wooden table, provided it is sufficiently secure that the hive cannot be knocked off. It also needs to be strong enough to take the weight of the hive when the supers are full of honey.

THE FLOOR OR BOTTOM BOARD

This is a shallow wooden tray that protects the bottom of the hive from damp. This board also includes the entrance block, which is a piece of wood that can be used to adjust the size of the hole according to the time of year. There is also a sloping landing board (see below) in front of the entrance to give the bees somewhere to land.

THE COMPONENTS OF A HIVE

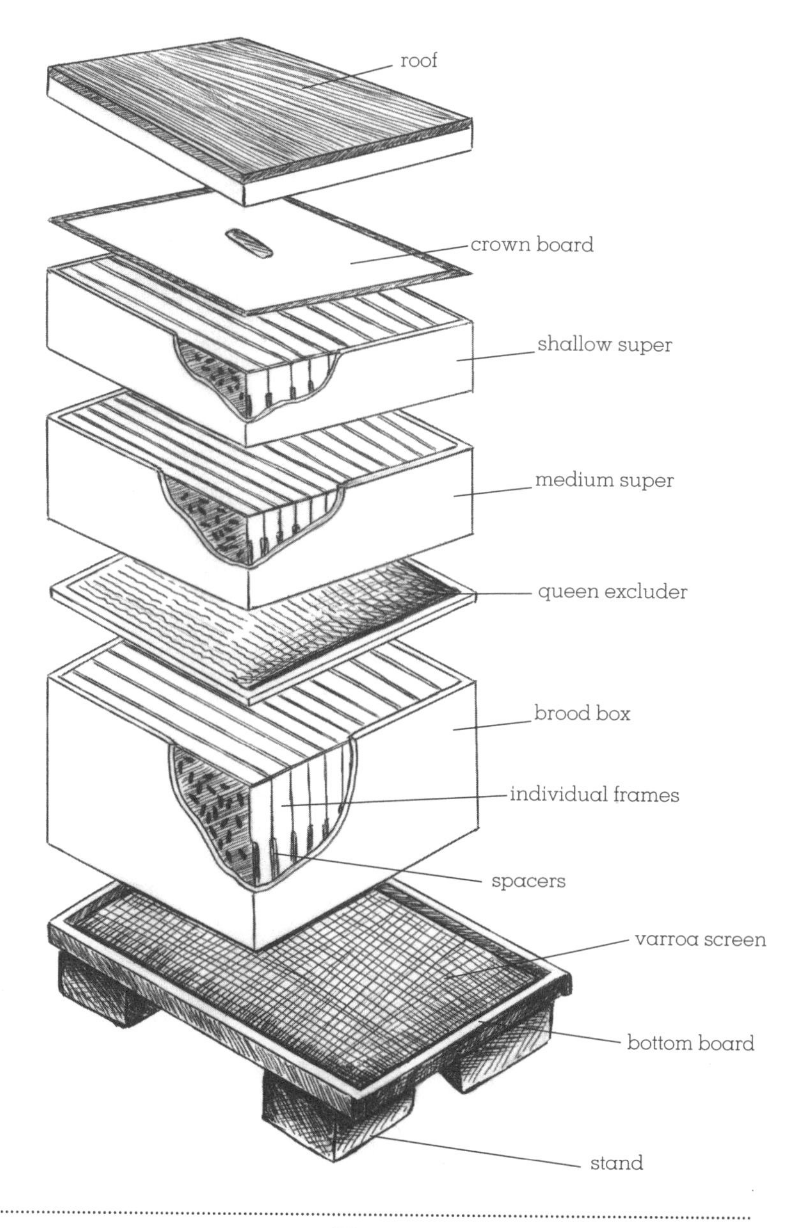

THE VARROA SCREEN

Between the bottom board and the brood box is a piece of mesh, designed in such a way to allow any varroa mites (see pages 138–139 for more detail) to drop through. These mites will die on the bottom board and can be counted regularly in order to examine the extent of infestation.

THE FRAMES AND SPACERS

These are the wooden frames that hang vertically inside the brood box and supers (see opposite page) and contain the wax foundation that stimulates the bees into making comb (see right). They come in either deep or shallow format, depending on whether they are intended for the brood box or the supers. The ones in the supers are those that will eventually be removed when they are full of honey. The ones in the brood box will contain the bee brood – worker, drone and queen. These frames need to be perfectly spaced, so it is advisable for the beginner to buy ones that have built-in spacers to avoid any problems within the hive.

FOUNDATION

A sheet of foundation is a man-made piece of wax

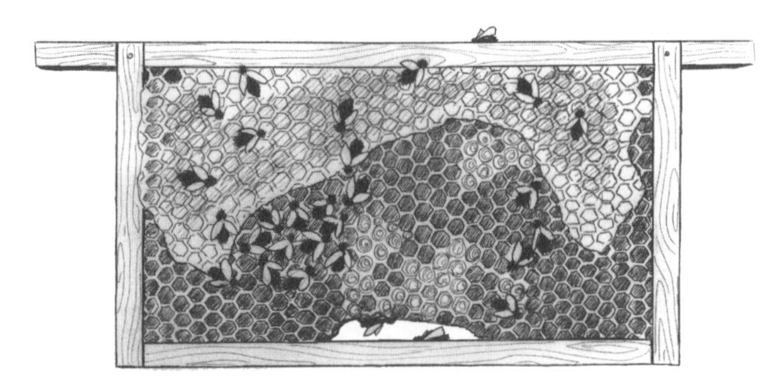

An individual frame showing some cells filled with honey.

that has been moulded in hexagonal cell shapes on both sides. This is used to encourage the bees to start making their own wax cells. Foundation can be purchased either wired or unwired; the former is much stronger and can be reused once the honey has been removed. Alternatively, if your aim is to store your honey as honeycomb (see the section on extraction on page 118) you will need to use the unwired type.

THE BROOD BOX

This is the box where the queen and her colony will spend the majority of their life and contains frames filled with wax foundation. It is the place where the queen lays her eggs and where the workers are constantly busy tending to her and her brood. This section of the hive also contains a certain amount of honey to feed the larvae. The bees will maintain this section of the hive at a constant temperature of 35°C (95°F) by vibrating their wings. It is important to remove the frames in the brood box only when the weather is calm and mild or you risk chilling the brood.

THE QUEEN EXCLUDER

To prevent the queen from spoiling the honeycomb in the super by laying eggs there, a queen excluder is added above the brood box to prevent her from entering the supers above. It is a simple mesh screen that allows worker bees to pass through to deposit the honey, but the holes are too small to allow the queen entrance.

THE SUPERS

The supers are boxes which contain the frames with foundation where your honey will be stored. Each super can hold 10–14 frames and they can be purchased in either shallow or medium depth. They are not as deep as the brood box because the frames can be very heavy when they are full of honey.

Once the bees start storing honey when their foraging flights are at a maximum in warm weather, you may need to add more supers to allow the bees extra storage space. Bees instinctively store as much honey as they can to make sure they have enough food to see them through the winter.

THE CROWN BOARD

The crown board, which is usually made of wood, is placed on top of the super boxes to retain heat within the hive. It has a hole in the centre which allows you to feed the colony when honey stores are low without disturbing the rest of the hive.

THE ROOF

The roof keeps rain out of the hive, but is ventilated to avoid any condensation building up inside. Roofs come in both flat and gabled styles; the flat ones are probably preferable for the beginner as they allow hive parts to be stacked on top of it during inspection. Pitched roofs, however, will help the rainwater drain away.

SKEPS

The skep was a basket woven from coils of grass or straw which preceded the modern hive. The basic skep had a single entrance in the bottom and no internal structure so the bees had to make their own honeycombs from scratch. The main problem with skeps was the entire basket had to be destroyed when collecting honey.

TOOLS YOU WILL NEED

CLEANLINESS

A well-maintained hive and clean tools are essential for ensuring the health of your bees and the quality of their honey. A contaminated comb or hive tool can easily infect a healthy colony of bees, so good apiary hygiene is paramount. All equipment needs to be sterilized between use and any sharp edges should be honed regularly. A blowtorch is a convenient way to sterilize wooden components – making sure you do not allow the wood to catch fire – and soda crystals make a mild disinfectant that can be used for washing tools, gloves, wooden frames and so on. Dissolve 500 g (1 lb 2 oz) of soda crystals in 4.5 litres (1 gallon) of water, but use with care as it is mildly corrosive.

There are a few basic pieces of equipment you will require when starting out. These are a hive tool, smoker, bee brush, queen catcher, queen marking cage, marking pens and protective clothing.

HIVE TOOL

This little metal gadget (illustrated above) is so invaluable to the beekeeper that you might like to have a few extra handy, just in case one goes missing.

The hive tool is used for levering out the frames from the beehive, scraping messy edges or getting rid of the build-up of propolis (see pages 124–125) or comb from hive parts.

THE SMOKER

This piece of equipment (illustrated above) is used to blow smoke into a hive just before inspection. It has the effect of calming the bees, which means they are less likely to sting. In the wild, smoke would be the forerunner to fire, so as soon as the bees sense the smoke they prepare to leave the hive. Before they go they need to consume as much honey as possible from the stores to last them until they find a new site, and this gorging has a soporific effect on the bees that makes them less likely to attack an intruder. The beekeeper aims to enter the hive after the bees are full of honey, but before they have the chance to leave the hive.

Choose a smoker that has a protective guard around the outside so that you do not have to wear gloves every time you pick it up. Buy a fairly large one so that you do not run out of smoke halfway through. Make sure the bellows are made of a strong, durable material as this is the first part of the smoker to wear out.

How to use a smoker

The smoker is designed to produce just enough cold smoke to give the bees the illusion that there is danger in the area. For this reason the fuel that you place inside the smoker should only smoulder, not burn – hay, grass, wood shavings, old egg boxes and cardboard rolled up into tubes all work well. It is a good idea to check that a material like cardboard or egg boxes has not been treated with a fire-resistant chemical before placing it in the smoker.

The smoker has a small hole in the base of the canister which is designed to allow only a small amount of air to enter. Attached to the side is a pair of bellows which are designed to boost the airflow so that the material inside smoulders more efficiently.

To use the smoker all you need is a gentle puff of smoke at the hive entrance five minutes before you intend to start your inspection. This gives the bees time to start sucking up the honey. As bees are inclined to move away from smoke, the smoker can also be used if you wish to remove them from a frame, but remember to use only a small amount as too much can make them cross. Practise using your smoker before you actually open the hive, as this is a very important part of beekeeping and keeping control of the colony.

BEE BRUSH

This is a soft brush with a reasonably long handle that is used for gently brushing the bees from a comb if you wish to get an unrestricted view or when you are removing the frames for extraction of honey.

QUEEN CATCHER CLIP

When the time comes to start marking the queen (see pages 47–48), you will need to invest in a queen catcher clip (illustrated above). It is a very simple piece of equipment that resembles a large bulldog clip. You simply open the catcher by pressing the sprung levers between your thumb and forefinger and then place it over the top of the queen. You can then gently close the catcher again without causing the queen any stress or harm. Do not worry if you catch a few workers at the same time – the holes in the catcher are large enough for them to escape but not large enough for the queen to squeeze through.

MARKING CAGE

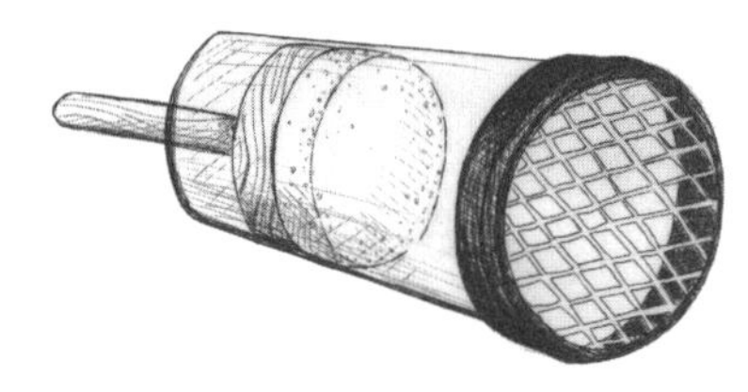

This is a small cage (illustrated above) that has been designed to hold the queen securely while you mark her head. It is a little round cylinder that has mesh at one end and a plunger at the other.

First, catch the queen with the queen catcher clip and transfer her into this marking tube. Once your queen is inside the cage, you gently ease the plunger up until the queen is in the top part against the mesh, when you can simply mark her through one of the holes. This is a great tool that is gentle on the queen and you will not need to worry about her escaping.

MARKING PENS

These marking pens are colour-coded, quick-drying and harmless. You will learn more about the colour coding in the section on queen marking on pages 47–48.

PROTECTIVE CLOTHING

As bees will see your invasion of their space as a threat, they will do whatever they can to protect themselves. To avoid getting stung it is advisable to have clothing that covers every part of your body (see page 41). Wear only white or light-coloured clothing as the bees will see bright colours as flowers and are more likely to land on you. It is also easier for you to see if you have any

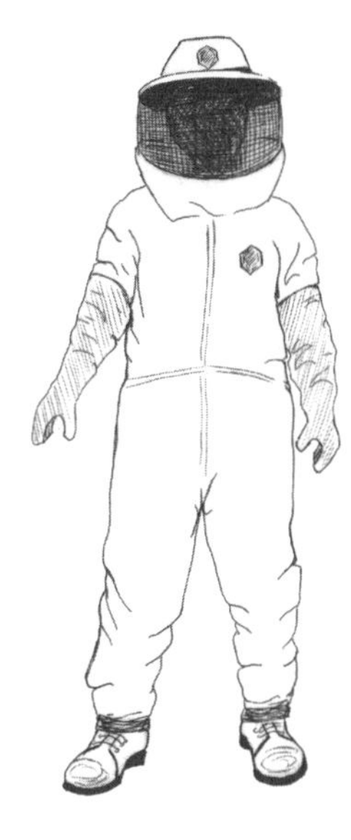

bees on you when you have finished your inspection – on darker colours they could easily be missed and then be squashed when you remove your suit.

The best outfit is the all-in-one bee suit with a veil and hoop built into the hood and elasticated cuffs to prevent the bees from going up your sleeves. This may be an expensive item but it will prove worth it in the long run. Bee suits come in both adult and children's sizes and some manufacturers will even offer to alter them for you if you cannot find a good fit.

The full body suit has elasticated cuffs and ankles and is designed to be worn over the top of a pair of strong leather boots or wellingtons, if you prefer. Tough leather gauntlet gloves are also a sound investment, although rubber washing-up gloves will protect from stings to some extent. Try to avoid wearing perfume or any other distinguishing smells, as this could also attract the bees. Make sure whatever you decide to wear provides you with plenty of ventilation, especially in hot weather, and that you are able to see clearly.

Bees are attracted to dark, secluded spaces, so do not have any loose folds, pockets, or any other areas of your clothing that looks inviting to a bee. If a bee does manage to crawl into a confined space where it is uncomfortable it will release its stinger.

FEEDERS

Although bees obviously would not depend upon human intervention in the wild, in domestic situations help is sometimes given to boost their food supplies at certain times of the year. This is particularly important once you have harvested the honey, as you have removed much of their vital winter store of food. During the cold winter months you will want to feed them with a strong solution of sugar and water, while in the warmer spring months they can get by with a weaker solution (see page 66). By feeding them in the spring you lead them to believe that the nectar flow has started for the year which kick-starts the queen into egg-laying.

Types of feeders

There are several ways of feeding bees, and one of the easiest is to use a rapid feeder. This is the same size as the brood box and has an opening in the bottom which allows the bees to come and go as they please. It is situated directly above the brood box and holds as much as 4.5 litres (1 gallon), which means that you do not have to constantly replenish it. It is a clean and simple method and can be topped up without removing the feeder each time.

The alternative type to use is the contact feeder. This is basically a plastic bucket with a tight-fitting roof that is upturned over the hole in the crown board. The roof has a hole that is covered in mesh, allowing the bees to enter the feeder, but the bucket forms a vacuum to prevent the liquid running out when it is turned upside down. This type of feeder comes in various sizes and can hold 1–4 litres (1¾–7 pints) of sugar syrup depending on your hive.

OBTAINING YOUR BEES

Now that you have bought your equipment, chosen your hive and sited it in a suitable position, your next question is where to obtain the bees that will live in it.

Late spring to early summer is the time to think about buying your bees and as a novice it is a good idea to either contact your local beekeeping association for reputable suppliers or follow up advertisements in beekeeping magazines. If you have already been on a course, you should have been given a list of local beekeepers in your area and it is well worth asking some of these for advice on the best place to obtain bees.

If you join an association and put your name down on the swarm list you may find that a local apiarist has just caught a swarm (see pages 91–95) you could take possession of – a free way of obtaining a colony of bees. They will arrive in either a nucleus or a package (see below). The disadvantage to swarms is that you have no idea what breed of bee you are obtaining and for the novice, it is a good idea to start with a breed that is known to be calm and non-aggressive.

A commercial bee farmer will be able to supply you with a colony complete with queen. However, you will need to plan this well in advance because very often demand is higher than supply, as the bee year is quite short.

HOW ARE THEY SUPPLIED?

There are three ways in which your bees will be supplied – as a nucleus (or nuc), a package or a full colony.

The nucleus

A small box that is really a temporary hive, a nucleus has a fixed floor, but the roof can be replaced with a wire mesh to prevent overheating when the bees are being transported. On one side is a small hole which has to be blocked up with either a piece of rolled-up newspaper, a cork, grass or something similar when the bees are being moved.

The nuc box will come complete with 10,000 worker bees, a laying queen and five frames containing brood, pollen and honey. The nucleus of bees is able to survive in the box for up to five days, but after this time the colony will run out of space as the queen will be busy laying. It is vital, therefore, that everything is set up before you receive delivery of your bees. To put the bees in their new home is a simple process that just involves transferring the five frames into the brood box in your hive and then leaving the bees to carry on their natural cycle without being

A nuc box is a convenient way of transporting a small colony of bees.

disturbed. This is an ideal way of obtaining bees as it gives you a strong start as opposed to buying a package.

Buying a package of bees

This is a cheaper way of buying bees than a nucleus as it does not include frames. It consists of 10,000 workers and a laying queen, which arrive in a screened box that contains some sugar syrup. Because the package of bees do not all come from the same colony the queen is separated in her own small mesh cage. This means the workers can get used to her smell and they can provide her with food through the holes in the mesh. It might take several days for the workers to accept the new queen as their own, but once they have done this they can survive together as a new colony. You can either pick up the package from the supplier or, if really necessary, it can be delivered by courier. Try to make sure that the journey they have to travel is as short as possible as you do not want to have a lot of stressed bees to deal with when they arrive.

The main disadvantage to buying a package of bees is that the colony will take several weeks to become established as the frames do not contain any developing brood. As workers have a very short life anyway – three weeks once they start foraging flights – the gap before the new bees are ready to take over means that your colony can get off to a slow start.

Buying a full colony

The most expensive option of all, this is the acquisition of a full colony of around 50,000 bees, including a laying queen, which will arrive in a working hive. This means it is ready to position in your garden and you can let the bees accustom themselves to their new location. The main disadvantage to this method is the risk of disease. Always check with your supplier that the hive has been sterilized before the colony is introduced to it.

The other disadvantage of a full colony for a novice is that you are being thrown in at the deep end and you have to learn very quickly how to cope with a large colony of bees. By buying your bees in a nucleus, you can get used to the gradual build-up and you will find it far easier to make your early inspections when there are not so many bees on the frame. You need to give yourself time to gain some confidence in the best way to handle your bees, and as you adjust to their behaviour and habits you will become more comfortable around them.

What breed?

If you are purchasing a nucleus or package from a reputable supplier rather than obtaining a free swarm you will be able to choose the breed of bee you want. The most common types in Europe and North America are the Italian bee (*Apis mellifera ligustica*) and the Carniolan bee (*Apis mellifera carnica*), both of which are bred for their non-aggressive nature.

Apart from their docile nature, the advantages in keeping the Italian honey bee are that the queen is a prolific breeder, they are resistant to diseases, excellent foragers, good comb builders and have less tendency to swarm than some other subspecies.

Carniolan bees, from Austria, Hungary and the Balkans, have a good sense of orientation and are not inclined to drift to another hive. They are less prone to rob honey than Italian bees, can survive winter better and can adapt quickly to any changes in the weather.

MARKING THE QUEEN

If you have decided to buy a nucleus of bees, it is a good idea to pay the supplier a little extra to mark the queen for you. Being able to see the queen is an important part of beekeeping as you will need to make sure that she has not left the colony or died. If she has been marked with a coloured spot on her thorax you will find her a lot quicker, which means your inspections will not be so intrusive.

It usually takes beekeepers quite a lot of experience and know-how to find the queen. To have a go yourself, find a frame where there are a lot of recently laid eggs. If you look closely, you may see a bee whose body is much longer than that of the other bees. She will be surrounded by a group of smaller bees, which will all be facing towards her. These bees are her attendants, whose sole aim is to look after her. The queen's job is simply to lay eggs, which she does at the rate of up to 2,000 a day, from early spring until the autumn when the colony becomes dormant.

INTERNATIONAL CODE

The international colour codes for marking queen bees are as follows:

Year ending in	Colour
0 or 5	Blue
1 or 6	White
2 or 7	Yellow
3 or 8	Red
4 or 9	Green

Marking a queen makes her more readily identifiable and using a colour code that indicates her age is particularly useful if you have several hives.

Marking the queen is a delicate operation that takes a little time to master. Some beekeepers wear thin rubber household gloves, while more experienced ones use their bare hands as they become more acquainted with their bees. The easiest way to catch a queen for marking is to use a queen catcher clip (see page 39) and then transferring her into a special marking cage (see page 40).

If you have not purchased a colony with a marked queen, you will need to buy some special paint or marker pens. With the queen safely trapped inside the marking cage, it will give you time to master the art of putting a dot of paint on her thorax. Once you are sure the paint is dry you can safely return her to the colony.

By using the international colour code (see page 47) you will be able to tell at a glance exactly what age your queen is. This is very important, especially if you are keeping several hives. Make sure you keep your records (see page 57) up to date with the ages and breeding condition of each queen.

The queen is much larger than either the drone or the worker, and a trained eye can usually pick her out quite quickly.

TRANSFERRING TO THE HIVE

Assuming you obtain your bees in a nucleus (see page 44) at the beginning of June, you will need to know how to transfer them to your hive. The very first thing you will need to do is place the nuc box close to the hive and leave it alone for the next 24 hours. This will give the bees time to calm down and acclimatize themselves to their new location.

As you do not want to encourage all the bees to leave the nuc box in one go, remove just part of the material that is blocking the entrance hole. If you leave only a small exit hole, the bees will not be able to get out so quickly and this will give them even more time to become accustomed to their new surroundings. They will need to reorientate themselves and adjust to the unfamiliar scents that are in your garden.

While they are still settling in, you can mix them their first meal of sugar syrup. This is very similar to the consistency of nectar and is consequently the perfect way to supplement the bees' natural diet.

The strength of the sugar syrup will vary according to the bees' demands. To introduce a nuc to a new hive you will need a weak solution – a ratio of 1:1 sugar to water. You will need to mix up at least 2 litres (3½ pints) to cater for 10,000 bees and make sure that all the sugar has dissolved before giving it to them.

If you observe a few simple instructions (see over) you should find it easy and trouble-free to introduce your bees to their home.

- Wait until the evening and light your smoker. Make sure you practise using your smoker beforehand.
- Put on your protective clothing and approach the nuc. As the temperature cools down the majority of the bees will have returned to the nuc box.
- Give a gentle puff of smoke close to the entrance hole of the nuc box and then move it away from the hive. This will give the bees time to calm down.
- Take the roof off the hive and place one empty brood box at the bottom of the hive.
- Bring the nuc box back to the front of the hive and, using your hive tool (see page 37) gently prise the top off the box. Do everything slowly and carefully so that the bees remain calm.
- Take the first frame out of the nuc box and place it in the centre of the brood box in the hive.
- One by one, remove the other frames from the nuc and place them in the brood box in exactly the same order as you removed them.
- If you spot the queen on one of the frames, take extra care as you do not want her to drop off or be injured.
- Once you have transferred all five frames there will still be a lot of bees left on the floor of the nuc box. Turn it upside down over the brood box and shake fairly vigorously until most have dropped into the new chamber. If any are left inside, lean the box against the entrance of the hive and they will soon follow.
- Making sure that you have the correct spacers in place, gently move the frames into the centre of the brood box and fill the spaces on either side with new frames filled with foundation. Depending on the size of your hive,

the brood box should take between 10 and 12 frames.

- Once you are happy that the frames are correctly spaced, place the crown board on top of the brood box.
- Before closing the lid, place the upturned contact feeder on top of the hole in the crown board and pour in your solution of sugar syrup. Do not worry if a few drips of syrup come out before the vacuum has built up inside the bucket.
- Because the roof will not fit over the top of the feeder, you will need to put an empty super on top of the brood box to accommodate the height of the feeder and then place the roof on top of that.
- Your final job of the day is to reduce the size of the hive entrance hole (see pages 76–77) to around 2.5 cm (1 in) so that there is not such a large area for the colony to defend.

Now you can leave the bees to get used to their new home. Lean the nuc box up against the hive for the first night so that any stragglers will eventually find their way to the colony.

In the morning you will probably be keen to get back to the hive to make sure that your bees have survived the night. Remove the nuc box, as this will now be empty, and spend some time watching the activity around the entrance of the hive. Once you become familiar with the behaviour of your bees, this activity can tell you a lot. Some bees will be hanging around the landing board just outside the entrance, while others will be making foraging flights to search for nectar and pollen. You may already see some returning with their pollen sacs full and bright yellow against their legs. Others will simply be flying around getting ready to find new foraging sites.

YOUR FIRST INSPECTION

You will probably be eager to open the hive and take a look at your new family of bees. However, you should be patient, since if you open the hive too often you may well find that the bees will leave to look for peace and quiet. Wait nine days before carrying out your first inspection. The reason for leaving it this long is because it will take the colony nine days before they cap over their first batch of eggs, an indication that it is safe to open the hive.

Make sure you choose a pleasantly warm day without any wind for your first inspection. Early afternoon is the perfect time as the majority of workers will be out foraging, which means there will be fewer bees in the hive. You must have everything ready before you start, as it is not advisable to walk away from the hive once it is open to the elements. Bees cannot cope with cool air for more than a short while, so your inspection has to be as brief as possible. You will need to wear your protective clothing and take with you any tools that may be required, such as your lit smoker, your hive tool and a clean, white cloth in case it is needed later on (see page 55).

Because there are many things to look for during your first inspection, you might like to ask an experienced beekeeper to accompany you and give you his or her valuable advice. It is a good idea to make a list of exactly what you are looking for so that you can tick off each job individually – in your initial excitement you might easily forget to check a few important items.

MAKING A START

The day has arrived when you are to carry out your first inspection and the weather is perfect. On your checklist of things to look for are:

- eggs
- larvae
- sealed brood
- queen
- queen cells
- pollen, honey and sealed honey.

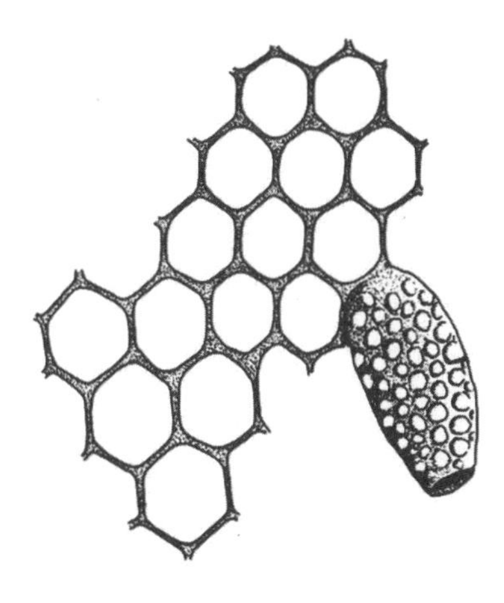

A queen cell is a cone of wax about the size of an acorn, usually hanging downwards at the bottom of the frame.

First, light your smoker. Put on your protective clothing and pick up the smoker and hive tool. Place the hive tool somewhere within reach – not in long grass as it is easy to lose – and give a gentle puff of smoke at the entrance of the hive. Wait a few minutes, then approach the hive from the rear; your shadow across the front of the entrance could make the bees angry.

Lift off the roof and place it on the ground upside down, first checking to see if there are any bees on it. Remember to use only gentle pressure and try not to knock the hive as you do not want to excite the bees inside. The roof may be a little difficult to remove as the bees may have used propolis to fill some gaps, but you should be able to ease it off without too much difficulty.

Next, remove the empty super you used to cover the feeder and place it next to the roof on the ground. Take the feeder out of the crown board and check whether there is any syrup left. If there is still some in it, the feeder can be returned to the bees at the end of your inspection if needed.

If you are a little nervous, you might like to give an extra puff of smoke through the hole in the crown board and wait a few more minutes before removing it. It will most certainly be stuck down with propolis, so you will need to run the sharp end of your hive tool around the edges to release the board. Work calmly, quietly and as fast as possible so the bees are not disturbed enough to become angry. Do not panic if you find a lot of bees on the underside of the crown board as you lift it off the hive – this is quite normal. Do not attempt to remove them; simply lean the board against the front of the hive, near the entrance hole, and the bees will find their own way back inside.

Inspect each frame

Your next task is to inspect each frame, one at a time. Using the curved end of your hive tool as a lever, gently exert pressure until the frame comes out of position. Start at one end and work methodically along the row, always replacing the frame in the same position that you took it from.

Do not be concerned if the ones on the outside are just empty frames of foundation – this is to be expected, as the bees will start their brood cells on the centre frames first. If the frame is empty, place it on the upturned roof so that you do not risk treading on it. If the second one is empty also, put it with the first to give yourself room to separate the other frames within the brood box. Before lifting any out for inspection make sure you have separated them all first, using your hive tool to prise them apart.

When you come across the first frame with bees on it, do not take it away from the brood box – simply hold it over the top so that any bees that drop off will fall back into the box. When you have finished, place the frame back in exactly the same place it came from. When you reach the middle frames you will notice the bees have been working hard since you introduced them from the nuc box. Things you might see are:

- cells full of liquid nectar
- cells containing variously coloured pollen
- stored honey cells covered with white wax
- sealed brood cells covered with a darker-coloured wax.

What you are looking for now is signs of brood at different stages of development. The larger larvae, which are white and C-shaped, are quite easy to spot. However, if you look closely you should be able to see smaller, rice-like grains at the bottom of some of the cells. The easiest way to see these is to hold the frame up so that the sun is behind you

WHAT TO DO IF YOUR BEES ARE ANGRY

You will quickly learn to differentiate between the normal buzz of active bees and the heightened sound of angry bees. If they sound threatening after you have opened the lid, simply cover the brood box with a white cloth and walk away from the hive. Puff yourself all over with the smoker to mask any smells that might be upsetting the bees and then return to the hive. Give an additional few puffs in the air just above the hive and hopefully by now the bees will have settled down.

and shines down on to the cells. The chances are that if you are holding a frame that contains eggs, the queen will also be present. It may take you a while to recognize her, but she is much larger than the workers or drones and you should be able to spot her longer body and legs. Take care with this frame as you do not want to knock the queen off – just place it gently back in the same position, facing the same way, and continue your inspection. By now you should be gaining confidence in handling your hive.

Finishing your inspection

Once you have checked all the frames and made sure they are back in their original positions, replace the frames that are lying on top of the upturned roof. Check the crown board is clean by scraping off any propolis or wax with your hive tool then gently place it back on top of the brood box, making sure you do not crush any bees that might be on the edges.

If the weather is warm and your bees have been foraging, you should have seen some stored pollen and nectar cells on the frames. This means you will not need to replace the feeder and you can simply put the roof on top of the crown board. If the colony is expanding nicely, it is now time to open the entrance by another 2.5 cm (1 in) so that the bees have more space to come and go as they please. You will not need to add a super until the frames in the brood box are full.

You can now leave the hive, content that your colony is doing well. Put out your smoker by placing a cork in the neck and before removing your suit get someone to check that there are no stray bees on you. Write the details of your visit in your notebook while they are fresh in your memory.

THE IMPORTANCE OF RECORDS

It cannot be stressed enough just how important it is to keep up-to-date records of your hives. Although you might think this routine is not necessary if you have only one hive, as the year progresses you will find that you refer to your records time and time again as an invaluable part of your beekeeping enterprise. By constantly looking back at your notes you will be able to check on how your colony is progressing and that way you will be able to quickly assess if the bees are in trouble.

After your first inspection you should have noted down the following points:

- the date and time of the visit
- the weather at the time of inspection
- the mood of the bees
- exact details of what you saw on the frames, such as good brood pattern in all stages; stored nectar and pollen; queen cells; queen in residence
- actions taken, such as marking the queen; feeding sugar syrup; adding or removing supers; signs of pests or diseases and any treatment given (see page 132–139).
- things to do at next inspection: add new supers (see pages 58–60); add new brood frames; return to check queen cells; take measures to prevent swarming (see pages 92–93).

You will need to keep a separate record for each hive as each colony can behave in a completely different way. There are no set standards for keeping bee records, so just use the method that works best for you.

WHEN TO ADD SUPERS

It will not be long before your new colony has outgrown its home, and to avoid the bees leaving to find a larger space it is up to the beekeeper to provide an extension. The bees will have been busy adding to the bare foundation in the brood box with their own wax so that the cells can hold brood, pollen and nectar, and you will need to start installing supers so that the colony can continue to grow unhindered. This is an important part of beekeeping as you want to give your bees enough room to expand so that you have a good supply of honey in the summer.

As soon as you see the brood box is becoming overcrowded and there are no spare frames, it is time to add a super above the brood box. There is no set time or date to do this and it will probably change each year, so you will have to assess the amount of space the bees have when you carry out your inspections. These need to be done regularly during the height of the season – approximately every seven days or so – as the situation can change from week to week. Remember that your bees dislike being disturbed, so do not increase the number of inspections beyond your weekly schedule.

It is not usually necessary to add a super after the initial inspection as the colony will not have outgrown its brood box, but as the weather warms up and more brood emerge, the colony will expand rapidly and need human intervention. If it does become overcrowded the bees will be inclined to swarm (see pages 91–93) and you certainly do not want this to happen with your first colony.

You will know that a super is definitely needed when your bees start to draw out the comb on the outside frames in the brood box. Drawn comb is where the bees have expanded the cells with their own wax to a depth of 12.7 mm (½ in). If you are uncertain it is better to be on the safe side and add a super anyway.

HOW DO I ADD A SUPER?

Adding a super is a quick and easy procedure but you will still need to light your smoker and put on your protective clothing. This is the first time you will use a queen excluder (see page 35) and you will also need your hive tool.

After you have calmed the bees with your smoker and waited for five minutes, remove the roof and crown board and then take your queen excluder and place it over the top of the brood box. This screen will prevent the queen from moving up into the super and laying eggs but will still allow the workers access to deposit their honey. If you find there are a lot of bees right at the top of the brood box, you can encourage them to move down by giving them a gentle puff of smoke.

You may need to use your hive tool to scrape off any propolis or spare wax that has built up on the top of brood box so that the queen excluder can sit flat without leaving gaps that the queen could squeeze through. As you place the excluder down, gently move it from side to side to encourage any stray bees to move away. Take a super with frames containing foundation and put it on top of the queen excluder, making sure that they face the same way as the ones in the brood box. Now replace the crown board and roof.

TOP OR BOTTOM SUPERING?

Learning when to add extra supers is quite an art, but you will quickly recognize the signs within the hive. A general rule is that if the bees have covered more than half the first super with honey cells, then it is time to add a second one. If the honey flow is prolific, it is not uncommon for bees to fill a complete super within a single week.

Experienced beekeepers use two types of supering – top and bottom. Top supering is probably the more common and is simply adding one super on top of another, as in the diagram above right.

If you wish to try bottom supering, then you always place the new empty super just above the brood box. This means you will have to remove any supers that are full of honey first. Position the new super and then replace partly full supers on the hive. Some beekeeepers believe this encourages the bees to move to the new empty super faster because the top ones contain some honey which acts a magnet for the bees.

TOP SUPERING

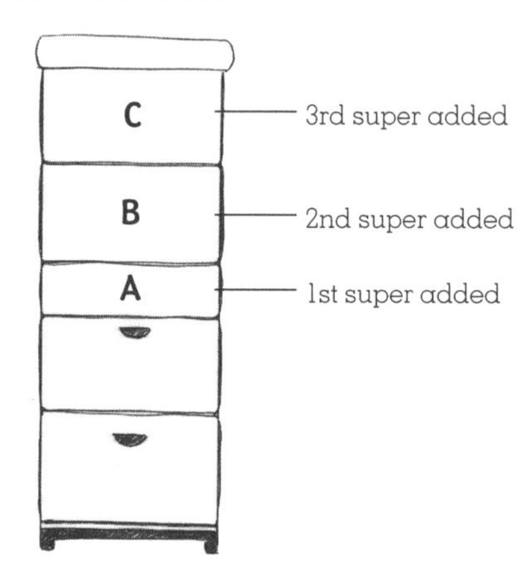

BOTTOM SUPERING

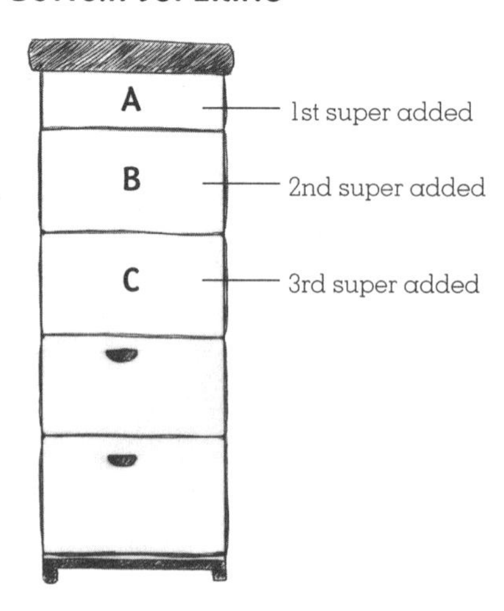

THE BEE GARDEN

To prevent the indispensable honey bee from going into further decline, it is vital to learn what to plant in the garden to attract them. Many of our more popular choices of garden flowers contain very little nectar or pollen and are not therefore considered beneficial to the bee garden. For example, double-headed flowers such as chrysanthemums, roses and dahlias put on a colourful display but provide very little pollen and no nectar. Because honey bees have short tongues compared to bumblebees, butterflies and moths it is far better to cultivate plants that are simple and open as they are the easiest for the honey bee to access.

Fruit trees, herbs and wild flowers all provide a plentiful supply, but we can help considerably by planting flowers that will provide the bees with food when supplies are short in early spring and late autumn.

Spring flowers are an essential source of nectar for bees, so make sure you have plenty in your garden. In winter, aconites, snowdrops and crocuses, along with shrubs such as Christmas box and winter honeysuckle, will help your bees to forage close to home. By April, they will be feeding from daffodils, forget-me-nots, tulips and hellebores.

They will also be drawn to shrubs and trees such as viburnum, currant, crab apple, flowering cherry, pussy willow and hawthorn. By growing at least some of these near the hive, you will be helping your bees to survive until they can start to fly further afield as the weather becomes warmer. The honey bee will travel up to 5 km (3 miles) to collect nectar and will visit around 50–100 flowers on each foraging flight. Bees can make as many as 24 trips in any one day.

If you have room in your garden, leaving part of it to run wild is a great way of making a good environment for bees. Clover, rosebay willowherb and brambles produce rich amounts of pollen. As autumn approaches, thistles, heathers, ivy and late-flowering crocuses can provide pollen later in the year.

A vegetable garden will also attract bees, so plant fruit trees, flowering currants, raspberry, blackberry and gooseberry bushes as well as beans, marrows and tomatoes. In between the vegetables you can add marigolds and sweet peas, all of which will encourage the bees and help with pollination on your plot.

Trees are not quite so important to rural bees, but urban bees will probably use these as their main source of food. Alders, poplars and hazels will flower at the beginning of the year, while later in the season sycamore, chestnut and acers will provide bees with a much-needed supply of food.

Although we can do everything possible to encourage bees to forage in our own gardens, when the bright yellow fields of oilseed rape are flowering in the countryside the bees will be drawn to it. Rape provides a vast quantity of nectar and pollen and the honey produced from this source is quick to crystallize and sets with a fine texture.

Bear in mind it is not the size of your garden that is important, it is what you plant and when. Even a few pots on a patio or windowsill can encourage bees.

PART 2

A BEEKEEPER'S CALENDAR

YOUR FIRST YEAR

Unlike most calendars, this book starts the beekeeper's year in June. Assuming you are new to your hobby, this is around the time you will have received your first nucleus. Although you will need to manage your hive throughout the year, there will be certain times when your intervention is limited as once the bees have formed their winter cluster (see page 79) your internal inspections will have to stop. Your main concern during the colder months is to check on the food stores and this is done by a process referred to as 'hefting' (see page 80).

As a beekeeper your main aim is to keep the bees happy, ensuring they have plenty of food and space and that the conditions are right to keep the colony healthy and strong – damp is a killer.

This section of the book is intended as a guideline to help you plan the year ahead after you have received your first colony of bees. It gives you tasks to carry out month by month and will assist in building your confidence when it comes to handling the bees. Regular inspections will teach you to quickly recognize the mood of the colony and spot whether there are any problems such as disease.

The calendar will also cover the problems of swarming (pages 91–93) and robbing (pages 90–91), which can be another major concern as this can introduce disease to your colony of bees. It will give you guidelines on what and when to feed, the way to cope when the nectar flow is at its height and how to identify the right time to start extracting the honey.

JUNE

You have just picked up your first nucleus of bees and you have made sure that everything is ready to receive them. You have bought a bee suit to protect yourself and have the correct tools to hand. Once you have allowed your bees to become accustomed to their new location, move the frames from the nucleus into the hive (see pages 49–51).

Because you are starting with a small colony which will not have had time to build up any stores, it will be necessary to feed them with sugar syrup (see page 66). Mix up the syrup, place the upturned feeder bucket into the crown board and close the hive. Now you will need to leave the bees alone for the first few days so that they can get on with their job of increasing the colony. Do not attempt to carry out your first inspection until nine days have elapsed.

It is essential to provide a supply of water, especially if it is a particularly dry June. Without water the bees will not be able to continue looking after their brood and storing honey. In hot weather, bees use water to help keep the colony cool; they collect it and spread it throughout the colony in tiny droplets. They then fan their wings to create a draught of air over the top of the water droplets. This causes the water to evaporate, which helps to bring down the temperature within the nesting areas. They also use water to dilute the honey to feed to the bee larvae. Place some water in a shallow dish near the entrance of the hive and replenish it as necessary.

June can be a difficult time for bees because of what is known as the 'June gap'. Spring flowers are largely over and summer-flowering annuals and perennials may not yet have started to bloom. You can help your bees through this period by planting shrubs such as cotoneaster and pyracantha, which can fill the gap by providing a source of nectar during this period.

First inspection

After your bees have been in their new hive for nine days, you can carry out your first inspection. This section recaps the things you should be looking out for:

- that your queen is healthy and is laying eggs
- that the brood is growing at different stages in the cells within the brood box
- that there are no visible signs of disease or pests
- that the floor of the hive is clean
- that there are signs the foragers have been out collecting nectar and pollen and some cells are being used for storage.

SUGAR SYRUP

Sometimes it is necessary to supplement the bees' stores. This can be done with either heavy sugar syrup or a light syrup. Heavy syrup is given to bees to help them prepare for winter, while the light syrup activates the bees in spring as it can simulate a nectar flow. Candy (see page 76) is fed to bees when the weather is too cold for them to cope with the amount of water in the syrups. Take care when feeding any of the above as it can excite the bees and disturb their winter cluster. To make the syrups, heat the mixture until all the sugar crystals have dissolved.

SPRING – Light syrup
1 kg (2¼ lb) white sugar to
1 litre (1¾ pints) water
AUTUMN – Heavy syrup
2 kg (1 lb 6 oz) white sugar to
1 litre (1¾ pints) water

HOW TO WORK SAFELY AROUND BEES

If you are new to beekeeping and you still feel a little nervous about working around your colony, follow this simple set of guidelines to minimize the risk of being stung or, indeed, causing injury to your bees.

- Remember not to wear perfume, aftershave, deodorant, hairspray or anything else that is highly scented. The smell will upset the bees and increase the risk of them becoming angry.
- If your hive is under or near a tree, check that there is nothing that is likely to fall on you or the hive before starting your inspection.
- If possible, have an experienced beekeeper with you when you carry out your first inspection.
- Make sure your smoker has plenty of fuel and that it is smoking well before embarking on your inspection. Do not leave the smoker anywhere that it is likely to start a fire, for example in dry grass.
- Check that your clothing is bee-proof and that there are no gaps or folds that bees can crawl into without you realizing.
- If you find that your bees are particularly aggressive on the day of inspection, increase the amount of smoke on both the bees and yourself, close the roof and walk away. Wait for a day when they are calmer.
- Do not stand directly in front of the hive entrance – always keep to one side. From time to time, give the entrance a puff of smoke.
- If you feel distressed at any point, calmly walk away from the situation. Bees are very sensitive to human behaviour, so if you cause them alarm they will punish you for it.

JULY

This is the month to look out for robber bees (see pages 90–91). As the name suggests, a robber bee is not a member of the colony but an invader attempting to steal the honey stores. The best way to spot if your hive is being robbed is to watch the entrance. If you see bees that are making hurried movements and look as though they are expecting to be attacked, you can be certain that these are robbers.

If your colony is strong, there should be no danger of robbing – it is only the weak ones that are vulnerable. Make sure you keep the area around the hive clean, do not allow sugar syrup to drip on the floor and never leave honeycombs exposed for any length of time as this will encourage robbers. If you do suspect robbing, partially close up the entrance with some dried grass so that the guard bees do not have such a large area to defend. A strong colony should be able to defend itself adequately, so if robbers have invaded your hive it is usually a sign that something is wrong.

Adding a super

Your bees have been installed in your hive for a month now, and it may be necessary to add a super (see pages 35–36). If the bees have worked their way across all the frames in the original brood box, it is time to add a super on top of it in order to give the bees room to expand and start new honey stores.

Before putting the super on, place a queen excluder (see page 35) on top of the brood box as this will stop the queen from going any further up the hive. You may need to add more supers as the bees start to fill the new frames so that as much honey as possible is made – some for you and some for the bees.

Adding foundation

The art of good hive management is knowing when to add new foundation, and this is a task that is very often neglected by novice beekeepers. With time, wax foundation can become damaged or start to discolour and even go black, which might indicate that there is disease present within the colony. This is a sure sign that you need to change your foundation.

By adding fresh foundation to the brood box and super, you will be encouraging the bees to create their new hexagonal cells and this will help to keep diseases at bay. You can buy foundation from a supplier of beekeeping equipment. If possible, buy organic so that you know it does not come with any added chemicals.

Foundation is made from clean beeswax, free of disease, which has been pressed into flat sheets. Each sheet is embossed with the imprint of the bottom of a cell which can be either worker or drone size.

Before beekeeping became more sophisticated, foundation was made from wax only. Most modern foundation has wires added for stability and to help prevent the centre falling out on a hot day or when the foundation is put into a honey extractor. However, if you wish to extract your honey as pure honeycomb, it is advisable to use unwired foundation instead. You will need to buy foundation that fits the size of your frames.

Alternatively, you could have a go at making foundation yourself. While the method might sound complicated, it is in fact relatively simple, although it does require both time and patience. For details of homemade foundation, see the instructions overleaf.

MAKING YOUR OWN FOUNDATION

The only specialized piece of equipment you will need is an embossed foundation roller, which can be purchased from a specialist supplier of beekeeping equipment. Alternatively, you may be able to borrow one from a fellow apiarist.

1. Melt some clean beeswax either in an electric wax melter or, carefully, on the top of your cooker.
2. Take some precut wooden boards of the correct dimensions to fit the size of your frame and then dip them in the melted wax.
3. Now cool the waxed boards in cold water.
4. Peel the sheets of wax away from both sides of the boards.
5. Run the wax sheets through the embossing roller. This will imprint the sheets with the hexagonal shape that the bees form naturally in their honeycomb.
6. Trim the wax sheets with scissors as necessary and fix them into a frame using the wedge bar and three nails.

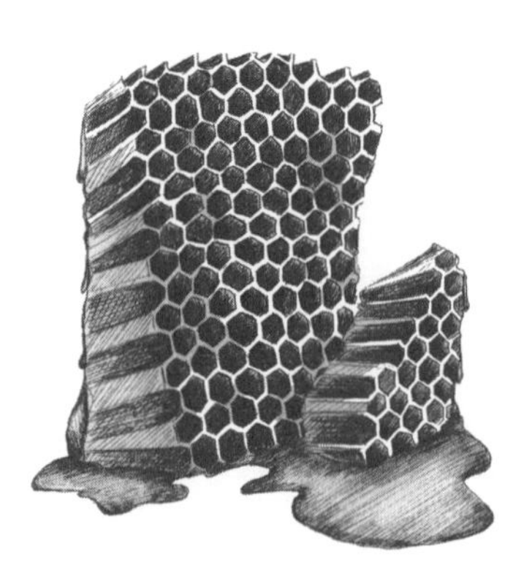

AUGUST

On your next inspection, check that the queen is still in residence and is continuing to lay. Inspect the frames in the super to see if the bees have filled more than half of them; if so, now is the time to add a new one. This means they will have plenty of honey to last them through the winter months. Do not expect to harvest any honey in your first year as the bees are going to require plenty in order to keep the colony strong and healthy.

Make sure that your bees are free of disease, particularly varroa mite, and refer to pages 132–139 if you are in any doubt. Continue to watch out for robber bees.

Make sure you are keeping your records up-to-date – you will need to refer to these next year so that you can see how your colony has expanded during your first season as a beekeeper.

The eviction of drones

Towards the end of the month you may start to notice that the drones (see pages 15–16) are being herded towards the margins of the hive. The reason for this is that as the end of summer approaches, the drones are driven towards the edge of the brood area and are eventually expelled from the hive altogether. However, as the climate is changing all over the world and autumn flowers such as goldenrod and wild aster are beginning to bloom earlier than their usual season, the behaviour of bees is also starting to change. The eviction of drones, which used to be at the end of summer to the beginning of autumn, is sometimes taking place later in the year.

But why are the drones evicted from the hive before winter? The reason is that they are quite distinct from the worker bees and do not share any of the same functions. The drone could be described as a highly specialized piece of machinery which is used for mating with the queen. They cannot help with the construction of the hive, nursing of the larvae or indeed gathering the nectar. The only time their jobs overlap is in controlling the temperature within the hive. The drone is also completely defenceless as he has no stinger, so he is no use in defending the hive against any predators.

The drone's role is complicated within the community because even though he is critical for the continuation of the species, he is not wanted in the community during the colder months because he would consume valuable winter stores without making any contribution to the general running of the hive. Because the drones are unable to feed themselves, keeping them supplied would also put extra pressure on the worker bees. Consequently, in the interests of the survival of the species, the drones are expelled.

BEE SPACE

In 1851, an American clergyman named Lorenzo Lorraine Langstroth discovered something called 'bee space'. He found that if you left just sufficient space between frames for two bees to pass one another – 10 mm (3/8 in) – they would fill this space with soft wax comb instead of the hard propolis. This discovery led to greatly increased yields of honey, more control over the honeycombs and subsequently healthier colonies. Today the most widely used commercial beehive is the Langstroth.

SEPTEMBER

The nights are getting cooler, the sun is losing some of its heat and the bees are making their first preparations for winter. Your priority this month is to check they have built up enough supplies of honey to see them through until spring. An average colony of bees will need as much as 18 kg (40 lb) of stores for the winter. You should be able to estimate how much they have according to the type of hive you are using: for example, each standard frame in a National hive holds just over 2 kg (5 lb) of honey.

Make your inspection in the first week of this month and check if there are any capped stores in the super or supers. If you notice that the super above the brood box is less than two-thirds full you will need to place a feeder containing heavy sugar syrup (see page 66) on the hive. Heat the mixture carefully until all the sugar crystals have dissolved completely. Use an upturned bucket feeder in the crown board and check a couple of times a week to make sure it does not need topping up. You will find that bees can empty a 4.5 litre (1 gallon) feeder in two days if the supply of nectar is short and as they will only take what they require, you do not need to worry about inadvertently overfeeding them. This type of feeding should be carried out until the first week in October. If they stop taking the syrup, you can remove the feeder.

Check the floor of your hive regularly for varroa mites (see overleaf) and treat appropriately.

Make sure you fill out your records with the exact amount of feeding, how many frames are full of stores, the state of the brood box, whether you have spotted the queen and the general health and wellbeing of the colony.

SUGAR SHAKE METHOD

The sugar shake method is a good way to determine the extent of varroa mite infestation. For this you will need:

- a small measuring cup
- a wide-mouthed glass jar
- a piece of light metal mesh
- icing sugar
- a plastic or cardboard box
- a backing sheet
- a light-coloured tray

1. Select a brood frame that is covered in bees.
2. Examine the frame carefully to see if the queen is present. If she is, return her to the hive.
3. Shake the bees from the frame into the plastic or cardboard box.
4. Try to get the bees to huddle in one corner by knocking the corner of the box on its side.
5. Scoop half a cup of the bees and place them into the wide-mouthed jar. Cover the jar with the mesh lid.
6. Add 2 tablespoons of icing sugar to the jar through the mesh lid.
7. Keeping the jar in an upright position, shake it from side to side to make sure that the bees are thoroughly coated in sugar.
8. Turn the jar upside down and shake over the top of the tray, moving the jar around so that the sugar and mites are spread evenly over the surface.
9. Count the number of mites on the tray, making a record of how many you have found, and repeat the process. Continue shaking and counting until no more mites are visible.
10. Tip the bees from the jar back into the hive and close the roof.
11. If you find more than 50 mites, treat your colony (see pages 138–139).
12. The best time to treat your bees is between harvesting honey and preparing your bees for winter.

COMBINING WEAK HIVES

For beekeepers who have more than one hive, autumn is the time to consider combining them if you notice one colony is becoming weak or has failed to build up enough supplies to get them through the winter. By combining a weak hive with a stronger hive you can maintain a good cluster size, which means the colony will have a better chance of survival during the winter. You will need to incorporate the bees from the hive with the weak queen into a hive that has a strong, healthy queen. The old, weak queen should be removed and destroyed and this should leave you with one strong colony.

The best way to combine two hives is as follows:

- Open the hive with the healthy queen, lay a sheet of newspaper over it and make some slits in the paper. The bees will gradually eat their way through, which gives them time to adjust to the combination. This means there will be less fighting and the bees from the weaker colony will come to accept their new queen.
- Destroy the weak queen and place the brood box with the bees from this colony on top of the newspaper-covered brood box.
- Leave undisturbed for at least two days, by which time the two colonies should have accepted each other.
- Remove the top brood box, first making sure you have brushed all the bees into the bottom brood box.
- Close the hive and leave the bees to unite and settle.

By spring you should have a healthy colony because there will be sufficient bees to maintain the winter cluster and a strong, healthy queen to fill the brood box as the weather starts to warm again.

OCTOBER

During this month you need to remove the syrup feeder as it could stimulate the bees to raise brood too early. If you want to make sure they have emergency rations you can always place some sugar candy, or fondant, inside the hive – see below for recipe.

RECIPE FOR SUGAR CANDY

To make the sugar candy, heat 2 kg (4½ lb) sugar in 600 ml (1 pint) water until it reaches 117°C (243°F) on a sugar thermometer. Then cool the pan in cold water, combining the syrup that is cooling on the edges of the pan with the warm syrup in the middle. As the temperature reaches 70°C (158°F) the mixture should start to go cloudy, immediately pour it into prepared pots before it sets hard.

To feed the candy to the bees, simply place 1 kg (2¼ lb) in a freezer bag and make several holes in the bottom. Place the bag, pierced side down, over one of the holes in the crown board on the hive and the bees will take this type of food when it is needed.

If you have been treating your hive for varroa mites (see pages 138–139), make sure you do not leave any chemical strips in the hive longer than is recommended by the manufacturers.

If you are storing any frames (full or empty), protect them against wax moth (see page 139).

Reducing entrance

October is the time to reduce the size of the entrance to the hive – though it is not just in the autumn that you should consider this. The dimensions of the entrance should be altered when necessary to cope with the size and strength of the colony within. Ideally, the entrance hole should be just large enough to avoid congestion during the busiest part of the day. In the height of summer when the nectar flow is at its height, the hole should be full size to allow for the continual coming and going of the bees on their foraging flights. However, a new nucleus or a weakened colony needs their entrance hole reduced so that the smaller number of bees can guard the hive successfully.

During winter when the bee traffic is considerably lighter, the entrance should be of a size to allow only one bee to pass through at a time. This means it will require fewer guards on duty so that the bees can do more productive jobs within the hive. If the entrance is not reduced and there are not enough bees on guard, it is an open invitation to neighbouring bees to rob your colony's winter stores.

The second reason for reducing the hole in October is to bar entry to mice and other rodents that will be looking for a sheltered place in which to take cover as the weather turns colder. A mouse can squeeze through a very small hole, which is why you will sometimes find wooden reducers referred to as 'mouse guards'. These are available to purchase from bee equipment suppliers and come in various sizes to suit the type of hive you have chosen. A determined mouse can chew through a piece of wood, therefore, the best type of mouse guard is a simple galvanized metal strip with two rows of holes in it, usually 9mm (1/3 in) in size, which are just large enough to allow the bees through. This can be easily attached to the floor of the hive entrance using ordinary drawing pins.

NOVEMBER

Unless the weather is unusually mild, your bees will have gone into their winter cluster by now (see opposite). As there will be no further need for thorough inspections, use the time to make repairs on any equipment that has been damaged, assemble new equipment for the next season and check that your stored frames are secure from wax moths (see page 139) and mice.

Examine the reduced entrance hole to check that it has not become blocked by dead bees. While frequent bee death is to be expected, if the living bees cannot exit the hive you may lose your whole colony.

Make sure your hive is secure against any bad weather and that it cannot be blown over in strong wind. Positioning it against a strong fence or a wall will give it added stability and shelter.

THE IMPORTANCE OF VENTILATION

The biggest threat to a colony of bees is damp, so the need to take adequate precautions against it occurring cannot be over-emphasized. In the winter months, good ventilation is very important. As the temperature drops at night, the cold night air mixes with the warmth of the hive and moisture can condense on the inside cover, forming drips which can fall onto the bees. If the bees get wet, they become chilled and have no means of warming themselves rapidly enough. Your hive must have a ventilation crack in the upper part to allow water vapour to escape – a 3 mm ($^1/_8$ in) crack at the front of the inner cover is sufficient. Anything larger than this can allow a robber bee or rain to enter the hive.

EXPLAINING THE WINTER CLUSTER

The term 'winter cluster' describes the way in which a bee colony survives the cold weather and absence of fresh supplies of nectar. With the eviction of the drones (see pages 71–72), the colony will have been reduced to about 10,000 bees – the size of your original nucleus. It is essential that there are this many bees in the colony at the start of the winter so that they can maintain a tight cluster around the queen. From late autumn to spring the queen will stop laying (or will lay only very few eggs) and it is up to the worker bees to keep her healthy enough to survive and produce the next season's progeny.

The bees huddle together in a large group, continually swapping places from the cold outer edges to the warmer centre of the cluster, where a temperature of 32–34°C (90–93°F) is maintained. During this time they keep up their strength by surviving on the stores they have built up in the summer and your role over the winter is to make sure that they have sufficient stores to last. Most of the bees now remaining in the colony were those born in the late autumn and they will survive until the spring, when a new batch of bees will take over the management of the hive. There will, of course, be some deaths, but these bodies will be dragged outside the hive, so do not be surprised if you see a few dead bees during the winter months.

You may also see a few bees flying around just outside the hive if the weather is relatively mild. These are 'cleansing' flights, when the bees take the chance to stretch their wings and defecate, since they need to keep the inside of the hive clean to avoid risk of contamination.

WHAT IS HEFTING?

Hefting is the term used to assess the amount of honey stores left inside a hive without opening the lid, and it is a necessity at this time of year. It is done by lifting first one corner of the hive and then the other. Experienced beekeepers are able to assess the weight by hand, but for the beginner it is easier to use a simple spring balance scale to lift up the hive and check each side in turn. The two figures are then added together to assess the overall weight. As a rough guide, one super full of honey should weigh approximately 13.5 kg (30 lb). Try to get a feel for the weight of the hive both when it is empty and when it is full. If you find you can tip it easily that would suggest the bees do not have sufficient stores; if it is hard to lift, this should mean that there is enough honey to see them safely through the winter.

METHOD OF STERILIZATION

As this is a quiet time in the beekeeper's calendar, you might like to take advantage of it to sterilize any equipment that you wish to use next year. Because hives are made of wood they provide perfect conditions for pests and diseases. By far the best way to sterilize them is to scorch them using a blowtorch.

Make sure you thoroughly scorch both the insides and outsides of all the hive parts, paying special attention to any nooks and crannies, where pests and diseases tend to proliferate. Do not burn the wood – you just need to give it a sufficient blast of heat to sterilize it. However, this is not something you will have to do in your first year unless you have purchased second-hand equipment.

DECEMBER

Christmas is here and it is time for you to take a well-earned rest from your duties as a beekeeper. Go to some local beekeeping association meetings or put a few bee books on your Christmas list. You might even like to make a list of the equipment you already have and what you still want to acquire. If you have enjoyed the first six months, you might consider getting a second hive and starting another colony. Two hives are not really much more work than one and you will be able to extract much more honey once the colonies are fully established.

Check that the bees have enough food stores left and that the hive is completely weatherproofed. If you see any problem areas, now is the time to fix them. Continue to check the floor of the hive for varroa mites and make sure your records are up to date.

BEE OR WASP?

Bees and wasps both belong to the order *Hymenoptera* and while they share some characteristics, e.g. they both have larvae which resemble maggots and stingers which inject venom, this is where the similarities end.

- the bodies of wasps are considerably thinner than those of bees;
- wasps build their nests horizontally, whereas the bees' nests are vertical;
- wasps can be aggressive and are able to sting more than once without dying;
- wasps die in the winter, whereas bees survive by clustering through the coldest season;
- bees have more hair on their body;
- bees have pollen sacs on their rear legs.

ALL ABOUT POLLEN

It may come as a surprise that some beekeepers are more interested in the pollen from their bees than in honey. Pollen has been regarded as a nutritious food supplement by humans for centuries as it is believed to contain many essential properties to maintain good health.

Pollen is the male gamete (sperm cells) of a plant, which reproduces by fertilizing the female gamete of the same species. This is where the bee comes in. The pollen grain adheres to the hairs that cover the bee's body, and the bee then uses its protruding tongue to brush the grains into the pollen baskets located on its hind legs. The bee then flies to a different flower and deposits some pollen grain, causing fertilization to take place. The bee then repeats this process, going from flower to flower spreading pollen grains as it goes, but will also save a good quantity of pollen in its baskets and take this back to the hive.

This is an interesting sight for the novice beekeeper to witness, as the collection of pollen will frequently look far too heavy for the bee to carry.

At the hive, the worker bees are ready to take the bounty. The returning bee passes her load to a worker bee which then finds a suitable cell to store the pollen. Honey is then added to the pollen and used to seal the cell until required for food. This mixture is known as 'bee bread' and is packed full of proteins, fats and lipids. Bee bread is fed to the young larvae and is crucial to the early stages of their development.

Pollen is believed to be just as good for human consumption and is considered a natural superfood. It is understood to contribute to a general sense of wellbeing and increase energy levels. To collect pollen from your bees you need a special device known as a 'pollen trap'. The traps available vary in appearance and the way in which they can be attached to the hive.

A pollen trap functions by directing the bee through a grid made of wire which causes some of the pollen to drop off the bee. Typically, you can expect to trap 60–80% of all pollen brought to the hive. All pollen traps are designed not to take so much pollen that there is insufficient left for the colony.

The pollen falls through the grid and lands in a drawer. Pollen is extremely perishable and so the drawer should be emptied daily, before it can decay. After collection, immediately transfer the pollen to paper bags and place them in the freezer overnight, which not only stops the pollen from deteriorating but also kills any wax moths and their eggs (see page 139) that may be present. You can leave the pollen in the freezer for storage, or dry it if you prefer – when the pollen is first collected it will contain a certain amount of moisture which will need to be removed to avoid deterioration. This can be done using pollen driers which are available from beekeeping suppliers, or you can simply air-dry it by spreading it on a porous surface in a well-ventilated room or greenhouse. Once the grains no longer stick together, it is sufficiently dry to store; use airtight containers made of metal or glass and keep them in a cool, dry place.

JANUARY

When bees are starting to raise their brood they need to extract more than just carbohydrates from their honey store to survive; protein is also very important to the health of the bees and this is obtained from pollen. If the temperature is above 10°C (50°F) you can feed them some pollen patties, but do not open the hive if it is colder than this as you will risk chilling the brood.

Providing pollen patties in mid-winter can not only maintain the health of the existing bees but also boost early brood production, which will help to replace the bees lost during the really cold weather. You can make pollen patties yourself, but if you need only a small amount it is best to buy them ready-made because bee pollen can carry AFB spores which can spread disease (see pages 133–134). Ready-made patties are made only from spore-free pollen. Allow 450–675 g (1–1½ lb) per hive.

Although it is generally not advisable to open your hive in winter, placing pollen patties in it can be done so quickly it does not break up the cluster. First, cut a 'V' shape in the paper around the patty and then peel it back to reveal the pollen dough. Open the hive and expose the top of the cluster. If it is in the lower box, you might have to remove the top box as the patty has to be placed just above the cluster, resting on the top of the frames.

Make sure you have your smoker lit, as you may need to encourage your bees to move down between the frames so that you do not risk squashing any of them. Check for varroa mite and if necessary treat with chemical strips (see pages 138–139). Hang two strips per hive for a maximum of 45 days.

PAINTING AND REPAIRING HIVES

During the winter months, make sure your spare equipment and hives are in good repair for the start of the new season. One of the most important jobs when repairing or repainting your hive is to check for any signs of pests or diseases. If the wood is badly infested, the best idea is to burn the entire hive. If the problem is only minor, sterilizing the hive before painting will suffice.

It is very important that all beekeeping equipment is kept in good condition. Keep an eye open for dry rot in the hive, which can affect the lid, bottom board and boxes. Depending on how serious the problem is, the wood can either be repaired or replaced. Dry rot can be detected by simply tapping the wood with a hammer – if it is present, the hammer will make an indentation in the wood. Frames are not easy to repair as wires can break, so it is probably better to melt down the wax foundation and replace the frames with new ones.

You will need to make sure there is no build up of propolis on any of the hive parts. This can be removed by scraping it off with your hive tool.

If you have had your hives for a while, you might like to put a preservative on the wood before repainting. If so, make sure the product you intend to use does not contain any insecticides and follow the manufacturer's instructions carefully. After preserving, leave plenty of time for the wood to dry before painting otherwise you might find the paint simply falls off. Use only non-toxic latex paint, as you do not wish to poison your bees. Choose a white or light-coloured paint so that it will reflect as much heat as possible.

FEBRUARY

February is the month when the bees start stirring within the hive. Things are starting to change within the cluster itself. The first cells are now being prepared so they are ready to hold the eggs that the queen is about to lay. If the winter has been really mild you might find she never actually stopped laying. As the brood is produced, the temperature inside the brood nest increases and this in itself will encourage more brood-rearing activity.

Woodpeckers

If you live in an area that has a green woodpecker population, February is the month you need to be particularly vigilant. Because the supply of insects is probably at its lowest, the beehive will be a great source of food for a green woodpecker (but not greater and lesser spotted woodpeckers).

There are several methods of protecting your hive against woodpeckers and which you choose is a matter of personal preference:

- Cover the hive in wire mesh.
- Wrap the hive in builders' polythene sheets.
- Hang old CDs above the hive.
- Paint your hive white; woodpeckers do not seem interested in pecking through paint.
- Use old fertilizer or animal feed sacks to cover the hive as these do not tear.

Ants

Ants often build nests in crevices within the beehive. The bees will not be bothered by their presence but you may find them an irritation you could do without. Treating them with an insect repellent runs a high risk of killing your bees too, so if you are really keen to deter them, stand the legs of your hive in cans filled with oil as this will stop the ants from climbing up the side of the hive.

Test for varroa mites and, if necessary, treat using two chemical strips per hive. Your treatment must be in the hive by 1 February otherwise you will not be able to install supers in April because you risk contaminating the honey. See pages 138–139 for more information on varroa.

Keep a check on the entrance to the hive to make sure that it has not become blocked, and that there is no build-up of dead bees in the area. If it is a really wet month, as is not uncommon in February, put a wedge under the back of the hive so that the floor slopes slightly forwards to allow any rainwater to drain away. Check that there are no holes in the hive that can allow water or wind in and that the roof is secure.

Gently heft the hive to check food weight (see page 80). If it feels light, place a block of sugar candy (see page 76) over the feedhole.

Now write down everything you have seen in your record book. Never leave anything to memory as these notes will come in very useful later in the year.

MARCH

By March, provided the weather is not too cold, your bees should be showing even more signs of early activity. By now the queen should be laying more eggs, which will be kept warm by the cluster, and the workers will start their cleansing flights. If the weather is really mild, you could see workers starting to bring in pollen from early-flowering plants such as the snowdrop, crocus and primrose.

Remove mouse guards

In the winter period when the bees were inactive you will have installed mouse guards to stop mice entering the hive, eating the honey stores and destroying the colony. As soon as spring comes the mouse guards will need to be removed, because the bees will need easy access for the collection of fresh pollen on which to raise their new brood. The removal of the guards ensures that the pollen will not become dislodged as the bees try to squeeze through a small space.

Check entrance

If you have not done so already, clear any vegetation away from around the entrance to the hive. If the weather is cold do not disturb your bees, but do check activity around the

entrance. Dead bees, bee excreta deposits (little sticky yellow spots on the outside of the hive) or crawling bees may indicate disease. Dead larvae show that the colony is running short of food. If you see bees fighting, this is a sign that robber bees (see pages 90–91) are operating, while a mass of bees coming out of the hive in a cluster means that the hive is overcrowded and the colony may be on the verge of swarming (see pages 91–93).

Superficial examination

Weather conditions are still changeable in early spring, so it is best not to disturb your bees unless it is absolutely necessary. This means you will have to observe the behaviour of your bees and other insects around the hive, checking for signs of disease and overcrowding.

The presence of a column of ants going into the hive may indicate that the colony has perished over the winter months. To check, put your ear to the hive and give it a good knock; if you do not hear buzzing coming from inside, you should lift the roof and investigate further.

Now is also a good time to remove any debris from the bottom board and check for the presence and if so, the extent, of varroa mites.

Continue to pay close attention to the weight of the hive by hefting (see page 80), especially if the weather is starting to warm up. The colony will now be growing quickly and the bees' consumption of food will be increasing rapidly. If they have already consumed the sugar candy block, replace it with a new one. You might like to consider feeding them with a light sugar syrup (see page 66) towards the end of the month if the weather continues to be mild.

ROBBER BEES

Colonies are at risk of attack from external predators throughout the year. Stronger colonies will seek out weaker ones, force their way in and plunder honey supplies. These trespassers are known as 'robber bees' and they can be the downfall of your colony. You may notice them buzzing around in the spring before the main nectar flow gets under way. If you see any of the following signs, there is a good chance your hive has been targeted:

- Dead bees in front of the hive
- Bees trying to enter the hive through splits or holes
- Bees at the entrance of the hive swaying back and forth
- Loud buzzing and the aggressive nature of the bees outside the hive
- Bees on guard inside the hive being more active than usual
- A large group of bees lurking outside the hive first thing in the morning
- Wax particles at the hive entrance
- Bees entering and leaving quickly.

There are steps you can take to avoid attracting robbers and lessen your chances of your colony being attacked, but first and foremost you need to have bred a strong colony. Robber bees will generally not attack strong colonies as they are able to defend themselves, having sufficient numbers to place guards at the entrance to the hive.

Not only will they deplete your colony's food stores, these raiders are also detrimental to the health and wellbeing of your bees. Robber bees can spread diseases such as AFB or EFB (see pages 133–135) as they can pass germs from one

You should be able to spot robber bees as they will be buzzing excitedly around the hive.

hive to another during their attacks. They can also cause stress within the colony, which in itself can lead to poor health and also the possibility of swarming.

The best advice is to reduce the size of the entrance as this will help the bees protect their hive by giving them a smaller space to defend. In summary, however, the long-term way to protect your hives from robber bees is to make sure your colony is strong and that you check your hives on a regular basis.

SWARMING

As the spring months advance, the beekeeper's job is to keep a balance between a thriving colony and an overpopulated one. Bees that do not have adequate food supplies, or enough room in the hive, will prepare to swarm. This is their way of reproducing a new colony by means of the queen and her workers departing the hive in a large group. The colony may also swarm when the queen is dying or becoming ill, or there are outbreaks of infection within the colony. The bees will leave in order to make a stronger colony in a new location.

Queen cells

Another cause of swarming is the emergence of a new queen. Look out for 'queen cells' on the outer edges of the comb, resembling small pods (see page 53). The queen will lay her eggs in these cells, and when they are capped she will be ready to leave with half the workers to form a new colony. When she is gone, a virgin queen will emerge to take over the colony, stinging her rivals to death. Once outside, the swarm will wait nearby while the scout bees look for a new nest. The scouts will then return and perform a dance, indicating what they have found, until the swarm decide which is the best place.

Although this is a completely natural process, it is not very helpful to the beekeeper, since the colony will be so depleted that it will not make much honey the following season. Therefore, beekeepers have devised various methods to stop this from happening.

Preventing swarming

There are many ways of preventing bees from swarming. Some of these are not very easy for the novice beekeeper, and may need to be performed by someone more experienced – for example, operations such as finding and destroying queen cells, or clipping the wings of the queen bee so that she does not fly away, taking her worker bees with her.

Others are quite simple, and involve good day-to-day management of your hive, such as feeding your bees adequately, especially in early spring, and adding a further brood box to make sure the colony has plenty of room.

If you already have more than one brood box you can also use the reversing method, swapping the positions of the upper and lower brood boxes, since the colony will have moved to the top of the hive during winter. Throughout the season, as the bees move up into the higher boxes, you can keep up this process of reversal. The queen will not go down to a lower brood box to lay eggs – if the upper brood box is full she will either stop laying or swarm. Putting an empty brood box from below on top gives her space to move upwards.

Another way of providing more room in the hive is to add supers on top of the brood boxes, so that the bees will have additional storage space for honey. The first box should be filled with comb, to help the bees with their honey production early in the season.

Catching a swarm

A swarm is a fantastic sight to behold – a black cloud of maybe 25,000 bees – but the noise can be quite alarming to onlookers. Although the thought of catching a swarm will probably be a frightening one, if you are with someone who knows what they are doing and everyone remains calm it is not as difficult as it sounds. There are specific individuals within beekeeping associations who have experience in this job and it is possible to put your name down on a list if you want to be the recipient of one of these swarms to build a new colony in your apiary. If you are lucky enough that the swarm is in your vicinity you can collect it and boost your apiary for nothing. However, this can be a risk as you do not know what breed the bees are and they may carry disease.

If bees are planning to swarm they will generally leave the hive in the morning and fly around in a rather haphazard manner for about 30 minutes before finding a temporary place to rest. They all gather in the same place and form a dense black mass, very often on a tree branch or fence post. To protect the queen they huddle in a shape that resembles that of a rugby ball, waiting for their scouts to find a new location. The swarm should be non-aggressive because they will have gorged on honey before deserting the hive, leaving them in a fairly soporific state.

This is where you can take advantage of the fact they are all in one place. Put on your bee suit, light your smoker and have a sturdy box handy. It is advisable to have someone with you as

EQUIPMENT FOR CATCHING A SWARM

- A sturdy container such as a bee skep or a cardboard box.
- A piece of net curtain or sheet to cover the box.
- Secateurs or pruning saw for cutting branches.
- Protective clothing.
- A smoker and a lighter or matches.
- A ladder if the swarm is high up.
- Brood box, frames, floor and roof ready to take the swarm.
- A white sheet for transferring the bees to the hive.
- A wooden board measuring 1.2 m × 60 cm (4 × 2 ft).

it is not really a job that can be attempted single-handed. Your aim is to dislodge the swarm from its temporary resting place into the box.

If the swarm is within reach it might be possible to give the branch a fairly vigorous shake so they drop into the box that your assistant is holding underneath. Alternatively, you could cut the branch with a pair of sturdy secateurs. If the swarm is on a fence or a gatepost you might need to use a bee brush to try to get them to move, first encouraging them with a gentle puff of smoke.

Once you have caught the swarm, cover the box with a sheet or net curtain and take them to their new location.

The easiest way to transfer your bees from the box into the hive you have prepared for their arrival is to place a piece of board up against the landing board and allow it to slope to the ground. Now cover the board with a piece of white sheet, making sure that it touches the entrance but does not block it. Turn the box containing the bees upside down and shake them all off onto the sheet. They will automatically crawl up the sheet, attracted by the dark entrance to the hive. Once the queen has entered the hive, the bees' natural instinct is to follow her inside. This can take up to an hour, but it is a natural process and one that fools the bees into believing they found the new home on their own. By using this natural method your bees will also settle much quicker in the new hive and quickly build up in the brood box.

Be prepared to place a super on this hive within the first week or so as this new colony will expand rapidly, since the number of bees is much larger than that of a nucleus.

APRIL

By now the bees will be busy collecting nectar and pollen from the abundance of flowers. Spring flowers are an essential source of nectar for bees, so make sure you have plenty in your garden. In late winter to early spring, aconites, snowdrops and crocuses, along with shrubs such as Christmas box and winter honeysuckle, will help your bees to forage close to home. By April, they will be feeding from daffodils and forget-me-nots.

They will also be drawn to shrubs and trees such as viburnum, currant, crab apple, flowering cherry, pussy willow and hawthorn. By having at least some of these near the hive, you will be helping your bees to survive until the weather becomes warmer and they can start to fly further afield.

It is also a good idea to check the entrance to the hive this month for the appearance of old pollen pellets. They will be small and grey and should crumble when crushed between two fingers. This is

COAXING BEES

Coax bees into early activity by planting the following in your garden:

Bluebells

Crab apple trees

Crocuses

Currant bushes

Flowering cherry

Hawthorn

Hazel

Heather

Pussy willow

Rosemary

Snowdrops

Viburnum

Winter aconite

Yew

a good indication that the bees are building up their brood nest. It is quite natural for these pellets to become dislodged when the bees use the entrance to the hive, and it demonstrates that there is plenty of pollen in the area.

Detailed inspection

When you make your first spring inspection, be careful not to take too long, or you may chill your new brood. If the weather is still cold simply lift the roof and remove the crown board for a couple of minutes. If you can see bees on at least four or five frames, that indicates that the colony has survived and is doing well. If the weather is warm and still, you can safely carry out your first spring detailed inspection. The main things to check are:

- that your queen is healthy, and is laying eggs. She should be fairly easy to spot as there are fewer bees in the colony with few, if any, drones
- that the brood is growing at different stages in the cells
- that there are no signs of disease or pests
- that the floor is clean
- that food supplies are adequate
- that the colony has started to grow bigger (you can do this by counting the frames they are utilizing)
- that the hive is not over-crowded (check for signs of swarming – see pages 91–93).

First, wearing your suit and carrying your smoker,

approach the beehive from the side, and gently lift the lid. If it is stuck, you will need to use your hive tool. If the bees are very active, use your smoker to calm them down. Next, take out the feeder frame at the end of the box. This will give you room to remove each of the frames, separating them with your hive tool.

In the centre, you will find the brood frames. Hold each of these up to the light. There should be a single egg resembling a tiny grain of rice in each cell, and the queen should be on one of the brood frames.

Now take a look at the brood cells. In a healthy colony, the eggs will be at the base of each cell, in a milky substance. You will also see white curled larvae. Each cell should be capped by wax. Check for the presence of pollen and honey around the brood area, and top up the feeder with sugar syrup.

Spring cleaning

Just like a house, the hive needs freshening up after a long winter. Wait for the warmest part of the day so that the foragers will be out and there will be fewer bees in the hive. Make sure you have a new brood box with you that contains frames filled with foundation, a new crown board and a feeder containing heavy sugar syrup (see page 66):

- Light your smoker, put on your protective gear and puff at the entrance to the hive. Wait for five minutes before opening the hive.
- Open the lid, remove the crown board and any supers and what you should see are your bees occupying at least four or five frames in the brood box – these might not always be in the centre.
- The frames that have bees on them are the ones that contain fresh brood. The

unoccupied frames are likely to be mostly empty and are the ones that should be replaced.

- Go to the frames that are furthest away from the bees first. Examine them to see if there are any capped brood present. If there are only a few cells of capped honey and no brood cells you can remove this frame from the hive, putting it to one side.
- Keep checking the frames in this way until you find one that has pollen on one side of it. This will indicate that there are some new brood cells on the opposite side. If you can see the larvae in the cells, put the frame back in the brood box in exactly the same place.
- Now go to the other end of the brood box and check the frames in the same way until you come across another frame with pollen and brood present. This should leave you with a total of about five frames with pollen, brood and bees.
- Move these five frames to the middle of the box. On either side place a frame that contains plenty of capped honey cells which will leave you with seven frames in the original brood box.
- Take your new brood box full of foundation and place it on top of the original brood box, making sure that the frames are lined up above the ones below.
- Place a feeder full of heavy sugar syrup over the hole in the crown board, allowing a couple of drips of sugar to fall on to the hive, giving a signal to the bees that there is food above.
- Close the hive and leave the bees to make their way up into the new brood box. As the bees start drawing out the new foundation, the queen

should naturally migrate to the new brood box. This is the bees' own way of making sure of the health of their colony, leaving the old used comb behind.

Leave the hive alone for about a week, making sure you top up the feeder with syrup if needed. On another sunny day, open the hive and see if the queen has moved up into the new brood box and has started laying new brood. If she has, place a queen excluder between the old brood box and the new one so that you do not risk the queen moving back down the hive. If you are unlucky and she has not moved into the upper brood box, try puffing a little smoke at the entrance to see if you can coax her to move up. If this does not work you will need to catch her and physically move her into the top brood box. She will quickly be followed by her attendants and she will start her new brood in the new brood box.

It will take 21 days for all of the brood to hatch in the lower chamber and move up to follow the queen. After this period it is safe to remove the old brood box and frames. Give the hive a new floor, entrance block, varroa screen and crown board and close up the hive. This is the end of the spring clean.

Sanitize the old brood box and put it into storage for future use.

Remember to keep an eye on the new hive, making regular inspections as the colony will be expanding rapidly at this time of year. Be ready to put a super on top of the brood box when more than half of the frames within it are full.

KILLER BEES

The Africanized honey bee, or killer bee as it is commonly known, is not a natural phenomenon but an accident of science – a cross between African and European honey bees that eventually bred with native colonies in the United States. They were originally bred to create a species that could thrive in the tropics and were introduced to Brazil in 1957. Since then they have continued to spread and now inhabit territories from Argentina to the southern United States.

Contrary to popular belief, their danger lies not in the ferocity of their sting (they are no more venomous than other honey bees) but in their sheer numbers. Most species of bee will send only a few guards out to chase you away from the hive if you disturb them. Killer bees will send almost the whole hive after you and will continue to chase you over a considerable distance. They particularly dislike a loud sound or a strong perfume. They are not inherently vicious, but more sensitive than European honey bees and quicker to defend their hives.

People have reported being stung hundreds or even thousands of times. While about 500 stings could be enough to kill a small child, the average adult can withstand approximately 1,000 stings provided they are not allergic to the venom. If you are unlucky enough to live in an area where killer bees are known to live, it is best to avoid any holes in the ground where you can see bees buzzing around, as this type of bee loves to build nests in such places.

MAY

By May you should be carrying out thorough and regular inspections of the brood box to make sure the queen is doing her job. After your spring clean any old combs should have been removed as they harbour disease, but if there are any left, take them out now. Look for any signs that the hive is becoming congested and add supers as needed.

Any chemical strips (see page 139) to control varroa should be removed before the bees start building up their stocks of honey in the supers.

Bait hives

You might also like to consider setting up a 'bait' hive to catch swarms that are looking for new homes – hopefully not your own bees! Several days before a colony is preparing to swarm, scout bees are sent out to find a new home. As the scouts return to the colony they perform the waggle dance (see pages 19–21), which is their way of reporting what they have found. The colony will interpret the dance movements and then decide whether it is worthwhile swarming. By setting up a spare or 'bait' hive, you can coax the bees into moving into a location of your choice.

First, they are looking for a space that is large enough for them to make a nest and will also give them room to store honey to get them through the winter. A brood box is just the size they need. Next, they will not want somewhere that has a large entrance that they will have difficulty defending, so make sure the entrance to the brood box is no more than 5 cm (2 in) square.

Bees prefer to go to a home that has been used previously by a colony, so installing old equipment that has not been cleaned of old propolis and wax will provide another source of attraction to the scout bees.

Next you need to decide where to place the bait hive. The best place is in a shady position as the bees do not want to work too hard trying to keep the queen and her brood cool on hot summer days. Cover the hole in the crown board, otherwise the new bees might be tempted to build comb in the roof.

If you see bees visiting the bait hive you can be pretty sure that there is a colony in your vicinity that is preparing to swarm. If you are lucky enough to catch a swarm, remove the old comb you have provided as quickly as possible and replace it with fresh foundation.

You will also need to check this new colony regularly for disease, as you cannot be certain where the bees came from and what they brought with them. Hopefully they will settle quickly in their new home and the queen will start laying a new brood.

Increasing the honey harvest

When flowers are in bloom the bees have more nectar and pollen sources and the 'honey flow' – the storing of surplus honey – will gain momentum. Your main aim at this time of year is to make sure your bees have enough room to store the honey.

A very effective way of doing this is to provide your bees with 'drawn comb'. This is foundation that has already been built out (or drawn) by bees, giving them less work to do.

When it comes to extracting the honey, you will want to take some for yourself, so encouraging them to produce large amounts is very important. You should also replace any supers you remove with new ones to allow the bees to continue their daily tasks.

REQUEENING

While the queen is the most important member of the colony, populating the hive by laying up to 2,000 eggs a day, there comes a time when her services are no longer up to scratch. Queen bees can live for up to five years, but they are really at their best in only their first and second years of life. You can measure how hard the queen is working by the number of emerging bees in the hive. After two years her sperm supply from her mating flight will be running out, and you will see a significant decline in the number of eggs she produces. If you leave her any longer, there is a good chance that she will stop fertilizing eggs, and the hive will fail as a consequence. Based on this, many beekeepers decide to requeen after two years. You can replace the old queen by buying a new one, or you can simply wait for the bees to take care of things in their own way by supersedure (see page 18).

The most important factor in requeening is the successful introduction of the new queen to the hive. You will need to inspect the frames in your brood box to check that you have a proportion of healthy brood and at least three frames of bees. The hive also needs to be queenless; if you need to bring this about yourself, destroy the old queen at least six hours before you introduce the new one. It is probably best to wait until your new queen arrives before you do this just to cover yourself against being left without a queen for a while.

There must be a good source of carbohydrate in order to create the perfect environment for successful requeening, but if there is a steady nectar flow at the time it is not essential to feed the hive. Once you are happy that

The new queen will arrive by mail in a small cage with several attendants. One end will be blocked with a piece of sugar candy.

your hive is in a good position to accept the new queen you can start the procedure of preparing her for her new home. She will arrive in a small cage and you should put a drop each of water and honey on the top and place her in a dry, warm place until you are ready.

Before introducing the queen, try to release the bees attending her in the cage – do this in an enclosed area, as you do not want to lose your queen. It is not essential, so there is no need to worry if they do not all leave. If the new queen is not marked, you will need to mark her before introducing her to the colony (see pages 47–48).

Your last job is to gently press the queen cage with its occupant between two frames that contain hatching brood. This is to make sure that there are young bees close to her at all times to take care of the new queen. One side of the queen cage will be blocked by a piece of sugar candy which the nurse bees already in the hive will chew through in a couple of days. Once the exit is open the queen will crawl out and climb onto the brood frame.

Leave the hive alone for five days to give the queen time to settle down with her new colony and start laying.

EXTRACTING YOUR HONEY HARVEST IN THE SECOND YEAR

Let us now assume that you have had your bees for a year and that you are approaching your second summer. You should really be able to start reaping the rewards for your hard work at this point. In a good year you can expect at least 18–22 kg (40–49 lb) of honey from each hive, but in a bad year you may have very little, so you will have to be prepared to feed your bees sugar in times of dearth. However, if your bees have survived the winter and are building up into a strong colony, there is no reason why you should not get a good crop of honey in your second year. Check that more than half of the frames have capped honey cells and that you have all your equipment ready before you start. You will not want to be running back and forth to collect tools once your hands are sticky.

You should not assume that having got your first colony of bees through the winter and early spring you have now learned everything there is to know about the hobby. There is still a way to go yet and more experienced beekeepers will tell you that every year they learn something new. Continue to read books, talk to professionals and consult with your local association if you are in any doubt.

Beekeeping is not labour-intensive, and although regular inspections are necessary in the summer the amount of time you spend with the bees is probably as little as 40 hours a year. Honey harvesting is a little more time-consuming, but consider the rewards: you can sit back at the end of the year with pots of honey on your shelves, content that you have not only learnt more about your hobby, you are also helping to slow the decline of the honey bee.

PART 3

PRODUCTS OF THE HIVE

MAKING HONEY

One of the main reasons you have chosen to keep bees is probably to enjoy the sweet golden liquid they produce – the honey. But do you know exactly how bees actually make it?

All honey starts out as a watery, sweet solution called nectar which is high in carbohydrates. Nectar is a natural product found in the pollinating part of a plant. When animals such as bees, butterflies, moths and hummingbirds go to collect nectar, they have to reach into the base of the petals and as they pass down the flower to do so, their bodies rub against the stamens that carry pollen, which sticks to them. As they move from flower to flower, some of this pollen rubs off onto the next plant's stigma, which ensures the successful continuation of that species. Plants that are pollinated by insects generally have bright colours or strong smells to attract them into the flower.

The honey bee uses its straw-like tongue (proboscis) to suck the nectar out of the centre of the flower. This nectar is stored in its stomach while it carries it back to the hive. During that journey (often a duration of approximately half an hour) the nectar is combined with other enzymes and proteins produced by the bee which converts the nectar into a simple sugar.

When the bee reaches the hive it regurgitates the nectar and passes it over to one of the worker bees inside the hive for the next stage. By placing drops on her tongue, the 'house' bee performs a kind of chewing motion to combine the nectar with air. This process reduces

the water content of the nectar and can take up to 20 minutes to complete for each individual droplet.

Once the house bee is satisfied with the consistency of the droplet, she hangs it inside one of the hexagonal wax cells. She will continue processing each droplet in this manner until the cell is completely full. To reduce the water content even further for long-term storage, the bees fan their wings rapidly around the honey cells to suck water out of the honey and thus thicken it. Pure nectar contains 80 per cent water, while honey, at the end of the process, has only about 15–18 per cent water.

Once the honey has been brought to the correct consistency, the bees cap the cell with a layer of wax and leave it to to act as their food store during the winter period when other food sources are not available.

Although you will take some of their honey reserves for your own use, it must be borne in mind that honey is vital for the survival of the colony. It is fed to the developing larvae, and the foraging worker bees and the nursery bees need it for sustenance, too. The fanning of their wings to maintain the hive temperature drains a lot of their energy, so they must constantly eat honey to replenish it. During one year a single colony can consume as much as 120 kg (265 lb) of honey – so you can see why it is vital that you leave adequate supplies for the bees to survive and thus keep your own honey pots full.

HONEY STORES

One of the most common questions asked by novice beekeepers is how much honey they can expect. The answer to this question depends on the weather, how prolific the nectar flow was for that year, how you manage the hive, and the health of your colony. You can help your bees to store extra honey by controlling the number of supers you place on the hive. A rough estimate of the amount of honey you can expect in your second year is 18–22 kg (40–49 lb). A National hive that contains, for example, one super with ten frames should hold 10 kg (22 lb) of capped honey. This is a lot of honey, so you will need to make sure you have enough containers to hand when you come to extract it from the hive (see page 112). If your colony builds up rapidly in the first year, you might find you can extract a few frames then, but you should be prepared to wait until the second year.

There is no set time that your honey will be ready to extract as this will depend on the weather, the abundance of flowers, plants and trees in your area and the strength of the colony. As a general rule you should be able to take off a couple of supers by midsummer and you could possibly get a late harvest at the end of the summer too. Each hive will perform differently, so you will have to treat each one individually and act accordingly.

TASTE AND AROMA

The aroma, flavour and colour of the honey your bees produce will depend on the type of flower the nectar has been collected from. It is very hard to predict what your honey will taste like as you can never be certain where your bees have collected their nectar. This can be interesting in itself, as you can try to judge by means of your nose and palate which flowers the bees have mainly been visiting.

If nectar has been collected from a variety of flowers, the honey is known as 'polyfloral'. If the bees collect nectar from only one source, it is known as 'monofloral'. Honey varies in colour and is measured by the Pfund scale. This system determines which colour category honey is graded into: light, amber or dark. The Pfund scale does not indicate quality; however, it has been noted that darker honeys contain more minerals than lighter versions, being rich in chlorine, iron, magnesium, manganese, potassium, sodium and sulphur. Whatever the type your bees create, your finished product can be affected by both the climate and the environment in which your bees are kept.

In most countries, the majority of honey is from a mixture of sources, with the exception of heather and certain clover honeys. As a general rule, the darker the honey is the stronger the flavour will be.

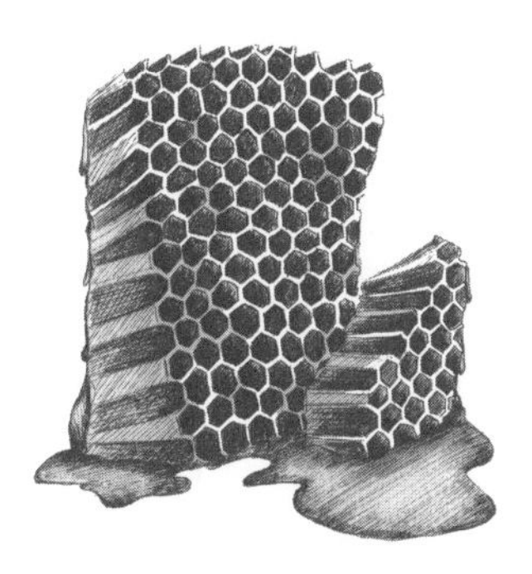

EXTRACTING THE HONEY

With the right planning and equipment, the process of extracting the honey can be a lot of fun – but however careful you may be, both you and the area in which you are working are going to get very sticky.

You will know that it is the right time to start harvesting your honey when the bees have capped their honey cells. This is an indication that the honey within those cells has matured sufficiently and is now of the correct consistency. If on inspection you find that three quarters of the cells have been capped, you can proceed to the extraction stage.

GETTING READY

Before removing the supers from the hive, you will need to have all the necessary equipment ready:

- an extractor
- an uncapping knife
- filters (or sieves)
- food-grade containers or buckets
- warm water
- cleaning cloths or sponges
- a spare super.

You will not be able to extract your honey out in the open, as the bees will try to reclaim it. You will need to find a sterile place that is free from dust – a kitchen or utility room is ideal. Make sure you carefully sweep, wash and disinfect the floor and work surfaces, leaving as much clear, hygienic space as possible. Have a bucket of warm water and a clean towel handy so that you can wash your sticky hands when necessary. Close all doors and windows to stop bees from being attracted to the smell of the honey. Ideally there should be more than one person to do the job, particularly as the full supers can be very heavy.

CONTROLLING THE BEES

Needless to say the bees will not be happy about you stealing their honey, so you need to find the best way of taking the supers without making them too angry. There are several methods you can try but by far the best for beginners is a bee escape board. Place the board, which has an escape hole in it with a one-way valve, between the brood boxes and the supers the day before you wish to remove the harvest. The bees will go through the hole in the escape board to spend the night in the brood box, but the valve does not allow them access again into the supers, which means your frames will be bee-free in the morning.

More experienced bee-keepers tend to use bee brushes to gently brush the bees off the combs. This method is not recommended for novices as the brush can quickly become clogged with honey and can also cause the bees to become aggressive.

THE EXTRACTING PROCESS

The basic process of extracting the honey from the frames is to use a hot knife to cut off the capping each side. Run the knife just under the surface of the wax capping, leaving the rest of the comb intact. Dip the knife in boiling water after each scraping to keep it clear of wax. Put the wax cappings into a container, as you can melt these later to make other items (see page 126).

After capping the frames, place two or more into an extractor – most of them hold up to four frames. The extractor is simply a spinning drum that forces the liquid honey out of the cells. There are quite a few different designs, but basically they all spin and the centrifugal force removes the honey. Radial extractors do both sides of the frame at once. If you use a tangential extractor you will have to turn the frames manually to get the honey out of both sides.

Once the extractor is full, put on the roof and turn the handle (or turn on the switch if it is electrically powered), starting slowly. Let some of the honey spin out before increasing the speed as this allows the comb to stay in one piece. Continue spinning for another few minutes until all the honey has come out of the comb. Remember that if you are using a tangential extractor you will need to rotate the frames and spin again.

When you are happy that you have removed as much of the liquid honey as possible, take out the frames and place them in the spare super. You can return these to the hive later for the bees to clean up.

Next the liquid honey in the base of the extractor needs to be filtered through a sieve into a clean bucket or any large container. Once this is done, cover the bucket with a clean cloth and leave the honey to settle for 24 hours. You will find that bubbles have settled on the surface; skim these off carefully. Your honey is now ready to pour into sterilized jars. When your jars are full, cap them immediately and store them in a cool, dark place. Put a label on each jar with details of which hive the honey was taken from and the date it was collected.

If you want to sell your honey you will have to adhere to the rules that apply to your country regarding hygiene, food standards and health and safety. In the quieter winter months you can read up on this subject or contact your local bee association.

MANUAL HONEY EXTRACTOR

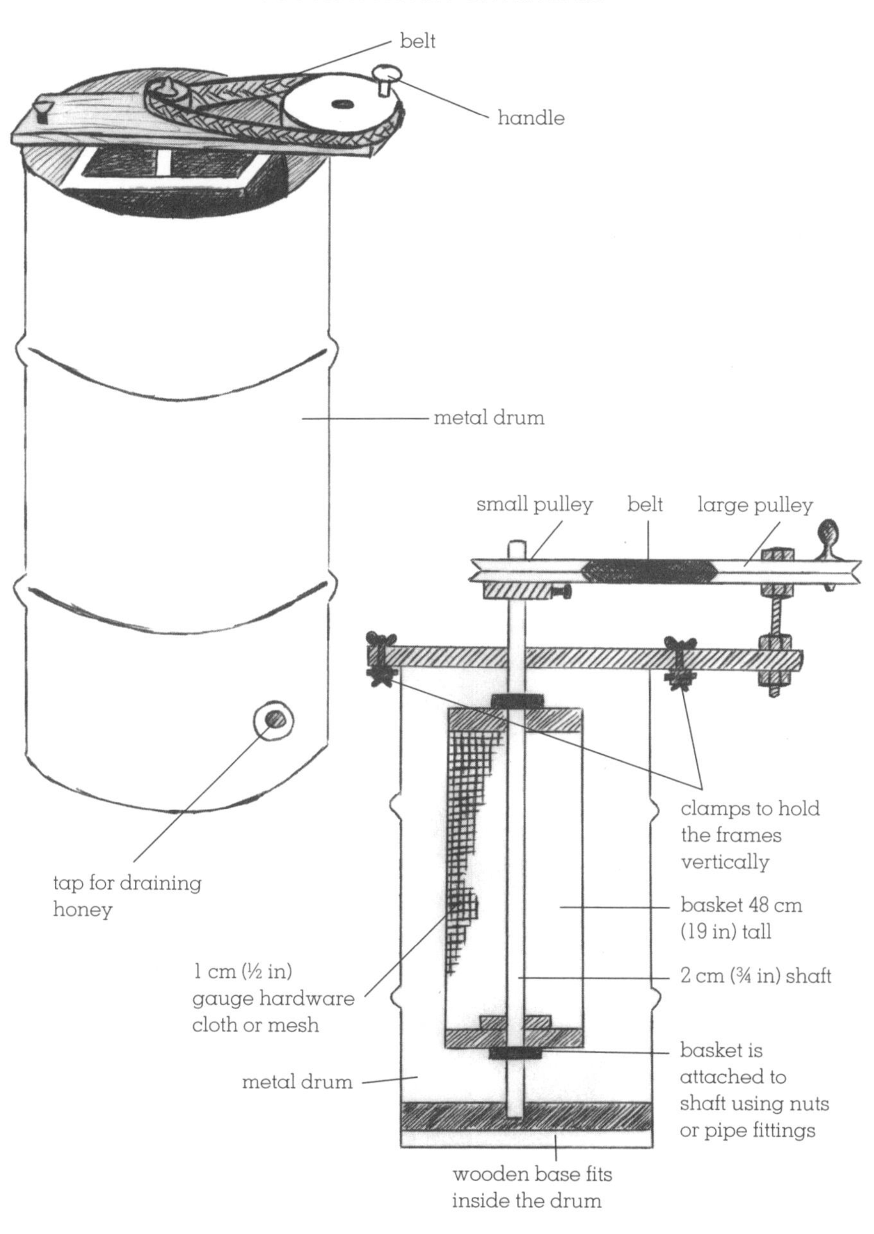

MANUAL EXTRACTION

If you do not have access to an extractor, it is possible to harvest the honey by hand by physically scraping the comb. This will take a lot longer and is far messier, but it produces good-quality honey nonetheless. The only problem with this method is that the honeycomb is totally destroyed, so it cannot be returned to the bees for cleaning. If honeycombs are replaced in the hive, the bees will have less work to do as instead of drawing out new comb they can start to store nectar straight away in comb that is already formed.

If you decide to use the manual method, you will need a large bucket to collect the honey and another one for the wax comb. You will also need a large spoon or scraper, a coarse sieve and a fine sieve for the final filter.

Start by placing the coarse sieve over the top of the honey bucket and hold the full frame over the top. Scrape the honey and wax straight off the frame into the sieve and then leave to filter through the mesh into the bucket. Once you are happy that as much of the honey has drained through as possible, put the wax comb into the other container and repeat with the other frames.

So that your honey is really clear, filter it again through a fine sieve or muslin into another clean honey bucket. Hopefully this will get rid of any minute particles of wax that passed through the coarse sieve.

Cover the bucket with a fine sieve and leave for a couple of days to allow any air bubbles to rise to the surface. Skim the surface and bottle.

Place the wet frames back in a super and return them to the hive for the bees to lick clean. Once they have been cleaned you can fill them with new foundation.

TESTING FOR PURITY

Honey is hygroscopic, which means that it naturally attracts moisture. If excess moisture were to get into the honey it could potentially ruin the batch, either by encouraging fermentation, or by ruining the taste.

A honey hydrometer can give you an accurate reading. Alternatively, there are simple methods for testing purity that you can carry out at home.

Home honey test

For these tests you need only some simple household items:

- a sample of honey
- a cotton wick
- a candle
- some matches
- a piece of blotting paper.

1. Put a sample of honey on a saucer and place it in the refrigerator. If the sample crystallizes it is impure.
2. Slowly pour half a cup of honey into a second cup. Pure honey spins clockwise as it is poured.
3. Dip a length of cotton wick into the honey sample. Allow the honey to dribble off the wick. Light the candle and hold it against the wet end of the wick. If the honey is pure, the wick will burn because there is a minimal water content. If the wick does not burn it means that there is a high water content present and the honey is impure.
4. Place a teaspoon of honey on a piece of blotting paper. If your honey is pure it will not be absorbed by the blotting paper. Impure honey will be.

COMB HONEY

Comb honey is, as the name suggests, the honey still in the comb and therefore the most natural form. When eating comb honey it is not necessary to remove the wax, as it is very thin and is pleasant when eaten with the honey. Some people prefer to extract all the honey from the comb, while others like to keep some in this natural form. All you need to do is to use a thin, sharp knife and cut out small squares from the comb. If you have been using wired comb you will have to remove the wire first before putting your squares into sterilized containers, so if you know you are going to keep honey in this form the best plan is to buy unwired foundation from the outset. The wires are intended to hold the comb firm while it is spun in the extractor, so they are not really necessary if you intend to only keep honey in comb form.

A good way of storing comb honey is to put smaller pieces of the comb in a sterilized jar and then fill the remainder of the jar with liquid honey. This is the best of both worlds and makes a nice present for friends or family.

If you have children, handling honeycomb is a great activity to encourage them to be involved in as they will see the honey in its purest state. They will be able to study the wax cells and appreciate how intricate and exact they are, and will also discover how the bees cap the individual cells with wax to stop the honey from running out. Of course, you will not want your children to be in their best clothes for this bit of biological education!

ROYAL JELLY

Royal jelly is a complex nutritious substance that is produced by the young nurse bees to provide food for the queen bee and her larvae at certain stages in their development. It is secreted from glands on the top of a bee's head and before it is fed to the larvae it is mixed with proteins and sugars in the bee's stomach.

The queen bee is fed solely on royal jelly and because it is so rich and nutritious she is able to reach maturity five days earlier than the worker bees. Also, when she is fully grown, her weight is double that of the workers. Her life expectancy is also longer – around five years as opposed to the workers, who are expected to survive only 35–40 days.

Because it has been proven to be such a beneficial food for bees, many people believe that it can offer the same benefits to humans. It is held to improve conditions including high cholesterol, high blood pressure and also many allergies. When studied, royal jelly has been found to contain considerable amounts of proteins, lipids, glucides, vitamins, hormones, enzymes, mineral substances and specific vital factors that act as biocatalysts in regenerating cells within the human body.

Royal jelly is the most profitable product produced in the hive and experienced beekeepers have devised specific ways of encouraging bees to produce higher quantities. The royal jelly is removed using small suction devices and it is either chilled or frozen so that it does not lose any of its goodness.

TRICKY HONEY

Most types of honey can be harvested easily. However, depending on the make-up of the honey you are producing, some types will not succumb to the pressure of the honey extractor. It is worth knowing about the composition of different honey types as you progress further into the hobby, but for now, examining three types – heather, ivy and oilseed rape – will give you an idea of the problems you could encounter in pursuit of your golden harvest.

HEATHER

To take advantage of flowering heather during the late summer and autumn, many beekeepers actually move their hives on to moors so that their bees will forage solely from this plant. Heather honey is thought to be one of the best types available, and has a clear, almost jelly-like consistency. However, when it comes to extraction, this honey can be a problem.

Heather honey is thixotropic, meaning that it has a reduced viscosity when pressure is applied, and it will not come away when the extractor spins. If you are harvesting heather honey for liquid honey, you will need to press the combs once they are removed from their frames. This can be done by using an improvised tool made from a wooden block with needles embedded in it.

Rollers for this purpose can also be purchased. These are made of plastic and have sprung needles attached.

In areas where heather honey is made frequently and in large quantities, beekeeping associations have special presses for this task.

IVY

Ivy has a very distinctive and strong aroma that dramatically affects the honey derived from it. It may not be to everyone's liking, so it is worth smelling the flower itself before you try making any honey. This type is tricky to handle as it sets in the comb, sometimes even before the cell has been capped, so when you come to extract it you must do so very quickly, otherwise you are faced with having to cut it out of the comb and melting it back into liquid form.

OILSEED RAPE

Because of its high glucose content, oilseed rape honey granulates rapidly. Like heather honey, it will not come away from the comb when spun in an extractor. One way beekeepers deal with this is to sell the honey as honeycomb, though not all customers may be keen on eating beeswax. If you want to extract it as liquid you need to remove the supers from the hive as soon as they are full and sealed, then break the combs away from the supers and gently heat them. The wax will solidify on the surface and can be lifted off. Run the heated liquid through a filter and stir before pouring the honey into jars.

MANUKA HONEY

Manuka honey is made from the nectar gathered from the Manuka bush (*Leptospermum scoparium*), a wild shrub indigenous to New Zealand. Although your bees will not have access to manuka bushes, this honey is worth knowing about. While it is recognized that honey is generally very good for you, manuka honey in particular is packed full of healing properties such as:

- antibacterial
- antifungal
- anti-inflammatory
- antimicrobial
- antioxidant
- antispetic
- antiviral.

Manuka honey has a major difference from other honeys; it contains a plant-derived component called methylglyoxal, a compound which experts believe is the key to its success. In the case of other types of honey, their medicinal properties are lost after exposure to light or heat, or even once they come into contact with wound fluids. Manuka honey remains effective and is able to destroy bacteria that causes infection at the site of a wound. Besides aiding the recovery of wounds or abrasions, manuka honey has been known to effectively ease the symptoms of a range of ailments and conditions, for example:

- aches and pains
- acid reflux
- acne
- blisters
- burns
- cold sores
- digestive problems
- infections

- insect bites
- nail fungus
- psoriasis
- stomach ulcers
- sun burn
- wrinkles.

Perhaps the most surprising of the conditions that manuka honey has been claimed to treat is the appearance of acne and wrinkles. The antibacterial properties of manuka honey soothe blemishes and prevent new ones forming. Most anti-acne remedies are made of solutions that dry out your skin, whereas manuka honey is a natural alternative which does not do this, leaving the skin soft and smooth instead.

While the wrinkles that inevitably come with age cannot be permanently erased by any miracle product, your skin can be improved by the use of manuka honey. Full of vitamins and antioxidant properties which keep your skin hydrated, this honey is believed to improve the skin's elasticity and increase the production of collagen.

The best way to apply this very useful product to your skin is by creating a face mask, just as the ancient Egyptians did.

EGYPTIAN FACE MASK

Wash your face in warm water and pat dry. Mix 2 tablespoons of Manuka honey with 2 tablespoons of milk and apply to your face.
Lie down and relax as it dries into a mask – this should take around 10–15 minutes.
After this time, gently wash the mask away with warm water and see the benefits for yourself!

PROPOLIS

Propolis is a resin that bees collect from the buds of trees. They store the resin in the corbiculae (pollen baskets) on the back of their hind legs and then transport it back to the hive. Propolis is also known as 'bee glue' and is normally dark brown, although it can vary in appearance and colour, depending on its botanical source. In its pure form it is a sticky substance at and above room temperature, and becomes hard and brittle if stored at a low temperature.

Bees have many uses for propolis, including:

- reinforcing the structural stability of the hive
- reducing the entrances to keep unwanted visitors out
- sealing up small holes and cracks to keep the hive draught-free and to prevent diseases entering the hive
- keeping the entrances smooth and in good condition for the bee traffic
- lining the inside of brood cells before a queen lays her eggs to make a hygienic and waterproof unit for the new larvae
- sealing the bodies of mice and any other predators that are too large for the bees to eject from the hive. Left unsealed, the bodies would decay and potentially become a source of infection.

Bee glue for humans

It is not only the honeybee that makes use of propolis – bee glue has a long history of being employed for human purposes. In Ancient Egypt it was part of mummification rituals, while the Assyrians used it to heal wounds. During the Stone Age it acted as a glue to secure arrowheads, and much later it became a component in violin varnish. Today, propolis is a common ingredient in many everyday items such as creams, chewing gum, cosmetics, lozenges, toothpastes, soaps and ointments. Propolis is believed to be active against certain viruses and bacteria and is marketed as an alternative medicine; it is thought to aid in the treatment of inflammations, viral diseases, scalds and burns, and dental problems such as ulcers. It is also said to strengthen the immune system and to be very good for the heart.

HARVESTING PROPOLIS

There are two ways to harvest propolis:

Scraping the propolis off the woodwork

This method is hard work and will take a lot of time, but you can end up with a good amount of propolis. The downside here is that along with the propolis, you may also collect scrapings of wood.

Propolis screens

Using a propolis screen, also known as a propolis grid, you can encourage bees to deposit the bee glue in one place. The screens, or grids, are made of plastic dotted all over with punched-out holes 4–6 mm (1/8–1/4in) wide. The bees will seal up the holes with propolis and when it is full you can extract the screen.

Immediately place the screen in the freezer and leave until the propolis segments become frozen. When they are ready, flex the screen just as you might an ice cube tray and the propolis pellets will fall out. You can then put the screen back in the bee hive in order to encourage further cultivation of propolis.

BEESWAX

During the winter months there is little to do with regard to maintaining your bees, so it is worth making use of the time to work with another valuable commodity obtained from your bees – wax. There are several things you can do with pure, clean wax:

- make candles
- make polishes for wood and leather
- waterproof fabric
- strengthen thread for sewing
- treat cracked hooves on domestic animals
- make new foundation.

Do not waste any wax from old combs and cappings – these should be rendered down and made into blocks for later use.

The easiest way to render wax is to place some in a large stainless steel pan containing hot (not boiling) water. Keep the water simmering until all the wax has melted and then strain through a thin mesh into a clean container. The wax will float to the surface of the liquid and after several hours will become hard.

Remove the wax from the top of the liquid and throw the remainder away. Break the wax into small pieces and put it in another stainless steel pot to melt it down again. Remember not to allow it to boil, as wax is flammable. While the wax is melting, stretch a piece of muslin across the top of an old milk or fruit juice carton. Once it has melted, pour it slowly through the filter you have just made so that it drips into the carton. Filter again if you find there is a lot of debris still in the wax. You may have to change the filter a couple of times so that it does not become clogged. Do not be tempted to squeeze the muslin to extract the last drop of wax as this could result in your having unwanted particles in your finished product.

BEESWAX FURNITURE POLISH

60 ml (2 fl oz) liquid soap
250 ml (9 fl oz) very warm water
113 g (4 oz) beeswax
500 ml (16 fl oz) turpentine
60 ml (2 fl oz) pine oil

1. In a stainless steel pan, dissolve the soap in the warm water and then leave to cool.
2. Place a bowl containing the beeswax and turpentine over the top of a saucepan containing boiling water. Heat until the beeswax has melted and then stir. Remove from the heat and allow to cool.

THE PERFECT BEESWAX

To get the best result from your homemade beeswax, take note of the advice in the following points:

- Never use iron, zinc, brass or copper containers to melt the beeswax as these metals can discolour the finished product.
- Do not allow the wax to boil as this not only makes the finished product brittle but also creates a fire risk.
- Store blocks of rendered beeswax in a cool, dry place after wrapping them in paper or plastic. Protect the beeswax from wax moth.
- Wax is capable of absorbing chemicals within its proximity, so never store it near any pesticides as it could then harm your bees if you use it to make new foundation.

3. When both the mixtures are cool, mix together with a wooden stick or spoon, adding the pine oil.
4. Store in a clean container with a tight-fitting lid.

MAKING BEESWAX CANDLES

Scented candles make a wonderful present. Beeswax candles burn cleanly and have very little drip. They also last much longer than regular candles.

You will need

a sharp knife
beeswax in block form
a double boiler
thermometer
candle dye (optional)
a wooden spoon or stirrer
candle scent of your choice
candle wicks
candle moulds of your choice

Making the candle

1. Using a sharp knife, cut the beeswax block into small pieces and place them inside a double boiler. Use only a medium heat as you do not want the wax to boil or scorch.
2. Once the wax starts to melt, slowly stir it to even out the heat distribution. Now use your thermometer to check the temperature.
3. Once the wax reaches a temperature of 77°C (170°F) but no higher than 82°C (180°F) you can add your candle dye. Add a few drops at a time and quickly stir using a wooden spoon or stirrer until the desired colour is reached.
4. Now add your scent, but err on the cautious side as candle scents are usually very concentrated, so add a drop at a time. It is recommended that you use about 1 tablespoon of scent to each 450 g (1 lb) of beeswax.
5. Now place your prepared wick inside the candle mould. Most wicks come

with a small metal disc on the base to hold them in place. If yours does not have this, use a small amount of glue instead.

5. Next, slowly pour the wax into the candle mould, making sure that the temperature of the liquid is no higher than 70°C (158°F). Make sure the wick stays in the centre of the mould by holding the end with a pair of tweezers if necessary.
6. Cut the wick down to about 5 cm (2 in) above the level of the wax. Leave the wax to set before using.

WORDS OF WARNING

Never leave a burning candle in an unattended room. Use only scents and dyes that are intended for candle-making as other types may be water-based and could make your candle spit when it is lit. Do not leave a burning candle near hanging fabric such as a curtain that may move in a draught.

A BIT OF FUN

Here is something you can make to amuse children. Although the noise might drive you mad it is a little bit of fun and something they will enjoy.

Take an empty aluminium can and punch a hole in the centre using a nail and a hammer – obviously, you will need to do this yourself. Next, give the children a piece of string about 1 m (3¼ ft) long to cover with melted beeswax. They can then stick the string through the hole in the can and tie it to a small piece of wood – a cocktail stick or match is ideal. The string is then pulled back so that the wood rests against the inside of the can. To operate, hold the can in one hand and with the thumb and forefinger of the other hand, lightly pull downwards on the string at varying speeds. This makes different sounds according to your movements.

SOME USES FOR BEESWAX

Beeswax is used in many everyday products, including face creams, body moisturizers and furniture polish. Below are some interesting uses for this natural ingredient:

- Men who sport a large moustache have traditionally used beeswax to stiffen their facial hair into a particular desired shape.
- Beeswax is a great lubricant if you have old sash windows that are difficult to open and close.
- Copper and bronze items can be coated with a little melted beeswax to prevent them from becoming tarnished.
- Many dairy manufacturers still use wax to glaze over their cheese to prevent it from spoiling as they say that it does not taint the flavour of the cheese.
- Used on drawer runners, beeswax will make them slide smoothly.
- The surface of a tambourine is coated in beeswax.
- The hemp used for tuning bagpipes is first rubbed with a coating of beeswax.
- Certain children's sweets, particularly jellies, contain beeswax to give them a natural texture.
- Granite worktops will have a lovely glossy surface if regularly polished with melted beeswax and then buffed up with a clean cloth.
- In archery, the bow strings are covered in beeswax to reduce friction.

PART 4

PESTS AND DISEASES

PREVENTION AND CURE

Unfortunately it is a fact of life that bees are subject to certain pests and diseases. As a beginner you will not be able to spot all the signs, but this section will give you an idea of what to look out for and what action to take. Good hive management is paramount, so keeping the area around the hive clean and tidy and making sure that your colonies are strong should help them fight off many diseases naturally.

There are two different types of diseases to look out for – those that affect the adult bee and those that attack the brood. Some of the more serious ones may never affect your hives, but you should still know about them just in case.

GOOD HIVE MANAGEMENT

Studying your hive and making thorough notes should alert you to any changes that are taking place. It is not always easy to spot a problem, but examining the brood box should give you a good idea. If the queen is present and laying and the cells are forming an even pattern with a light brown capping, then generally you can conclude that there is no disease to worry about. Also watch the behaviour of the bees. Warning signs include bees hanging around the entrance, seeming lethargic; a lot of dead bees at the entrance; a decline in the production of honey; and unusual smells emanating from the hive, such as damp or mould.

Take the following preventative measures and you should ensure that your colonies are strong and are able to fight disease:

- Make sure you keep your apiary clean and tidy; do not discard any comb or propolis nearby or exchange combs between hives
- If you are buying second-hand equipment, make sure it is thoroughly sanitized before using
- Be careful not to damage or squash bees during your regular inspections
- Try to avoid robbing and drifting by taking care not to spill sugar syrup outside the hive
- If your bees die, make sure you close the hive so that no stray bees can get access to the honey
- If you are in any doubt, always ask advice from either an experienced beekeeper or the Bee Diseases Officer, who can be contacted through your local beekeeping association
- Learn to read the warning signs of when your colony is distressed.

DISEASES THAT AFFECT BROOD

American Foul Brood (AFB)

AFB is a very serious condition and you must contact the relevant government agency if it occurs in your hive. Your local beekeeping association will advise you on who to contact and your apiary may become subject to official control by means of a programme of apiary inspections.

The larvae of an infected brood usually die after the cell has been capped as the bacteria penetrates the gut wall and multiplies in the body tissues. You will notice an irregular pattern forming on the comb, with the infected cells becoming sunken and discoloured. The caps may also be punctured and look moist. If you tilt the comb towards the light you will notice the larvae have dried and formed brown scales.

The way to test for AFB is to insert a pointed matchstick

into the affected larva. If you see a brown mucus thread when you withdraw it, your brood are affected.

Combating AFB

- Immediately contact your local apiary inspector, who will inform you on the best form of treatment.
- Prevention is far better than cure, so make sure you practise hygienic hive management at all times.
- Requeen on a regular basis as a young, healthy queen will lay healthier brood.
- Select disease-resistant bees.
- Replace brood nest combs on a regular basis to help reduce the concentration of organisms that can lead to disease.
- As both AFB and EFB (see opposite page) are considered to be stress-related diseases, always take care when moving bees. Try to move a colony at night with an open entrance as this can help to reduce stress.
- Always check food stores, as a lack of nectar or pollen can create a nutritional imbalance and weaken the colony.

Chalkbrood

This is a fungal infection that usually appears in spring when the colonies are starting to expand. The fungal spores are eaten by the larvae and start to grow in their gut. The dead larvae will appear chalky white at the beginning but eventually become very hard and are no longer attached to the wall of the cell. The affected larvae will be removed by the house bees and can often be seen on the landing board. There are no chemicals available for treating chalkbrood, so destroy any comb that is badly affected.

Combating chalkbrood

- This disease can easily be

spread by the beekeeper using hive tools, so clean them thoroughly before and after use.

- If the colony is badly affected, requeen from a healthy colony.
- Make sure there are sufficient bees within the colony to control the temperature and humidity within the hive.
- Try to acquire a disease-resistant strain of bee when starting out.

European Foul Brood (EFB)

EFB, like AFB, is a notifiable disease and one that you need to be able to recognize quickly. This bacteria feeds on the food in the stomach of the larvae and eventually starves it to death. As the larvae will experience pain, you may find them in unnatural positions and their colour changes from pearly white to cream. They eventually become dry and form brown scales. The treatment for AFB is the same as for EFB.

Sacbrood

This is a virus that can usually be seen from May to early summer. It affects larvae once they have been sealed in their cells and will result in them turning from a pale white colour to a pale yellow. As the body dries up, the head will curl up and the larvae will lie on their backs in the bottom of the cell. Adult bees usually recognize the problem, uncap the cell and remove the affected larvae. The adult bees can become infected by eating contaminated pollen or by ingesting the fluid contained in the larvae. Although infected bees stop eating pollen and feeding larvae, sacbrood is usually transitory and is not considered a major problem.

Combating sacbrood

- If approximately 25 per cent

of the brood is infected, then remove and burn the comb.

- If a larger amount is affected, it is advisable to destroy the entire colony by spraying pesticide inside the hive and closing it up for at least 24 hours.
- If you wish to reuse any frames you can either soak them in a dilute disinfectant solution (1 capful to a bucket), or leave the frames for a few weeks as this virus becomes non-infectious after this period of time.

DISEASES THAT AFFECT THE ADULT BEE

Dysentery

This is a symptom that tells you there is something wrong within a hive. It is equivalent to diarrhoea in humans, the tell-tale signs being soiled frames and combs and concentrated spotting around the entrance to the hive. If the colony is badly infected you may also see dead bees lying around the entrance. Dysentery is more likely to affect a hive during prolonged periods of cold weather as the bees cannot take regular cleansing flights.

Combating dysentery

- There is no specific treatment available for dysentery and prevention is better than cure.
- Maintain strong colonies that show good hygienic traits within the hive.
- Keep everything you use in the hive clean.

Nosema

This disease is caused by a parasite which impairs the bee's digestion of pollen, consequently shortening its life. Once it is inside the gut, the parasite multiplies and starts to eat the bee from the inside. The bee will try to cleanse itself by defecating more often and when it is unable to leave the hive for cleansing flights, the area inside the hive and around the entrance can become soiled. Other bees try to clean up the mess and pick up the disease in the process.

Combating nosema

- The best defence for nosema is to make sure you have a strong colony before it goes into its winter cluster.
- Give them plenty of food stores to see them through the winter.
- If the queen is old, provide a young, healthy queen.
- There are several different chemicals available to treat nosema – Fumigilin being the most successful – but check with your local beekeeping organization or experienced apiarist for more details.

PESTS

Small hive beetle

This is a small, dark-coloured beetle that inhabits bee hives. Although it is easy to confuse the small beetle larvae with those of the wax moth, if you look closely you will see that the legs of the beetle are larger and more pronounced. If you do have an infestation you will see the beetles running across the combs to find hiding places when you open the hive. You might also see adult beetles under the top covers or on the bottom boards. Maintain strong colonies with a healthy queen, keep apiaries clear of all equipment not in use, extract honey as soon as it is removed and destroy any beetles you find.

Tracheal mites

These mites live in the breathing tubes of the adult bee. Because they usually attack in the winter months, the expanding brood is left unattended and dies. The signs to look out for are deformed wings and distended bodies, but because the mite is so small it can only be detected by microscopic examination.

Combating tracheal mites

- There is no effective treatment for tracheal mites, so the best advice is to keep your colonies strong and make sure you practise good hive management and cleanliness.
- Keep in touch with your local bee association to see if they have come up with any treatments.

Varroa mites

This is a parasitic mite that has now become endemic throughout most of the world. You will need to constantly monitor your colonies for levels of infestation and, if varroa is found, it is vital that you act quickly or you risk the collapse of the entire colony. The varroa mite is about the size of a pinhead and is visible to the naked eye. It attaches itself to the body of the adult bee and feeds on it by piercing the skin. It eventually weakens the bee and spreads harmful pathogens and viruses.

Combating varroa

- Bees need to be encouraged to groom themselves and many beekeepers like to use powders such as icing sugar or talcum sprinkled directly onto the bees. The powder does not harm the bees, but incites them to groom and, because the mites cannot cling on to the bee, the majority will become dislodged and drop to the floor of the hive.
- Use a screened board (see page 34) in the base of the hive so that when the mites drop through the mesh, they are unable to climb back up and reattach themselves.

- Buy chemical strips that, hung inside the hive, are a slow-release treatment to control the varroa mite.

Wax moths

While wax moths can be a nuisance, they can usually be kept under control by the beekeeper and the bees themselves. This moth likes to lay eggs in the dark corners of the hive and as they grow into caterpillars they will start to feed on the wax, pollen and honey stores and eventually on the bee larvae. If the colony is strong they will force the moths out of the hive, but if you see any moths or eggs of the wax moth, remove them straight away. Always check in cracks and corners when you carry out your regular inspections.

Wax moth can also be a nuisance in stored frames and combs, so either store them with some moth balls or crystals or place them in a freezer as these pests cannot survive the cold.

COLONY COLLAPSE DISORDER (CCD)

As yet there are no definitive answers to the question of why whole colonies of bees are suddenly collapsing in many countries. However, there are many theories, which include stress, malnutrition, genetically modified crops and overuse of antibiotics and pesticides.

If your bees are near a farm it is worth checking with the farmer if he is intending to use pesticides. Ask him to notify you before spraying so that it will give you time to close up your hive and cover it over during this time.

Always handle your bees calmly, and if you have to move them do it in the evening when the colony is quiet to avoid stress.

GLOSSARY

Abscond Not to be confused with swarming, this term is used when bees leave the hive because the conditions are not right within it.
AFB American Foul Brood, a viral disease affecting bees.
Apiary The place where bees and hives are kept.
Apiculture The science and study of keeping bees.
Apis mellifera The zoological name for the honey bee.

Bee glue See propolis.
Beehive The container used by a beekeeper in which to keep his or her colony.
Beeswax A substance secreted by the worker bee.
Bottom board The floor of the beehive.
Brace comb Random comb that connects hive parts.
Brood A general term for young bees including egg, larvae and pupae.
Brood box The part of the hive where the brood is raised and the queen is normally found.
Brood food A nutritious substance produced by the worker bees to feed both the brood and the queen.
Burr comb Comb that is not part of the main comb within the frame.

Capped brood Cells containing bee larvae that are fully enclosed by wax.
Caste The different classifications of bees.
Cell A hexagonal-shaped chamber that is part of the comb.
Chalk brood A fungal infection of the brood.
Cleansing flight The flight made by a bee to defecate after a long period inside the hive.
Cluster Either a mass of bees, or the huddling action taken by the bees during the winter.
Colony A family of bees living in a single unit.
Comb A set of hexagonal cells made of beeswax by the bees to store food and raise brood.

Drawn comb A comb which contains completed cells.
Drone A male bee.
Dysentery A disease in adult bees characterized by diarrhoea.

EFB European Foul Brood, a viral disease affecting bees.
Entrance reducer A device to limit the size of the entrance to the hive.
Extraction Removal of the honey from the comb.
Extractor A centrifugal device to remove honey from the comb.

Feral hive A nest of wild bees.
Food chamber The part of the hive used to store pollen, nectar and honey.
Foulbrood A generic term to describe a bacterial brood disease.
Foundation A thin sheet of wax or plastic to be used as a guide for bees to make new comb.
Frame A segment made of four pieces of wood that contains comb.

Granulate The process by which honey crystallizes or becomes solid.
Guard bee A bee that stays at the entrance to protect the hive from invaders.

Hive The home of an individual colony of bees.
Hive tool A tool used to remove frames and maintain the hive.
Honey flow The period of time when nectar is in abundance.
Honeycomb A comb that is filled with honey.

Italian bee *Apis mellifera ligustica*, a subspecies of non-aggressive bee native to Italy.

Killer bee A name given to Africanized honey bees.

Larva The second stage of development in the life cycle of a bee.
Laying worker An unfertilized female bee capable of laying drone eggs.

Marked queen A queen bee that has been marked with a coloured spot.

Nectar A food source found in flowers that is rich in carbohydrates.
Nosema A disease that affects the digestive system in bees.
Nuc, nuclei, nucleus A small colony of bees often used in queen rearing.
Nuptial flight The mating flight undertaken by a virgin queen.
Nurse bee A young bee that tends to the larvae.

Pheromone A chemical substance that influences behaviour in another member of the same species.
Pollen A powdery substance produced by the male part of a flower.
Propolis A sticky resinous material that bees collect from plants to strengthen their hives.
Pupa The last stage in a brood bee's development.

Queen A fertile female bee that is capable of producing offspring.
Queen excluder A mesh screen used to stop the queen from passing up into the supers.
Queen substance A pheromone produced by the queen bee.

Requeen The practice of replacing the queen with a younger, healthier one.
Robber bees Bees which will rob a hive of its honey or wax.
Royal jelly A rich substance produced by worker bees to feed the queen or a bee less than three days old.

Sacbrood A bee brood disease.
Scout A bee which is responsible for finding a new nesting site.
Solitary bee A species that does not live in a group.
Stinger The defence mechanism of a bee.
Supersedure A natural process in which a colony of bees replaces its queen with a new one.
Supering Adding extra hive pieces for the storage of honey.
Swarm A collection of bees that is seeking a new home.

Top bar The top part of a frame.
Tracheal mite A mite which causes weakness or death in honey bees.

Uncapping The removal of the wax cover of honey cells before extraction.

Varroa mite A mite which clings to the body of the bee and weakens the colony.
Varroa screen A screen positioned in the base of the hive to allow varroa mite to drop through.
Virgin queen A queen bee that has not been mated.

Wax A substance produced by the wax glands on the bee and used to build combs.
Wax scale A hardened piece of beeswax.
Wax moth A moth that can infest weak hives or stored hive boxes.
Winter cluster A tightly packed cluster of bees to maintain warmth during cold weather.
Worker A female bee that is responsible for foraging and maintaining the hive.

INDEX